ENERGY

FOR EIGHT BILLION PEOPLE

HOW FOSSIL FUEL DEPLETION IS CHANGING THE WORLD

By Irwin Wallman

To order additional copies, please contact us.
BookSurge, LLC
www.booksurge.com
1-866-308-6235

TABLE OF CONTENTS
ENERGY FOR EIGHT BILLION PEOPLE

PROLOGUE

FIFTY YEARS AGO one afternoon the author came home from elementary school, very upset and told his mother that he had just learned that the United States had only a 20 year reserve of oil. He knew that gasoline was used to power his dad's car and it was made from oil. Therefore, he concluded that in twenty years we would run out of oil and there would not be any gasoline for his dad's car. He did not understand there was an ongoing effort to discover and develop new sources for our oil. The explanation, we had an ongoing program to find new sources of oil satisfied his anxiety and he was able to enjoy those Sunday car rides into the country side. How many miles per gallon did that Buick consume? Gasoline was only 17 cents a gallon, so who cared? The thought that someday there may not be enough gasoline for the family car remained in a corner of his mind for his entire life.

PRIOR TO WORLD WAR II, there was plenty of gasoline for the family car. The price was in terms of the number of gallons for a dollar, seven or eight gallons for a dollar. Most of the oil for gasoline was from oil wells within the United States. There was no thought that someday we would be dependent upon our enemies for the oil that would become essential for our basic existence.

WORLD WAR II showed how critical gasoline was to fight the war. There was gasoline rationing. Our fighting forces ran on gasoline. It was an essential weapon for fighting the war. Gasoline as a weapon could be understood when visiting the museum at Normandy, France, the site where the Allied forces invaded Europe. It was interesting that it was necessary to build an entire port to unload the fighting men and materials necessary to fight the war in Europe. It was even more interesting to learn the Allied forces needed such huge amounts of gasoline, it was not practical to send the gasoline across the English Channel in 50 gallon drums. They laid a pipeline under the English Channel to provide enough gasoline to fight the war, an enormous undertaking. Our leaders realized gasoline was not only essential to the Allied forces, but also realized it was essential to our enemies, the Axis forces. One of the largest air raids of World War II was against the German oil fields at Ploesti in Romania. The air raid was not successful because the Germans had taken enormous precautions to protect the oil fields.

AFTER WORLD WAR II, oil was cheap and plentiful. We built gas guzzling cars and the Interstate Highway System, which allowed those gas guzzlers to burn more gasoline for commuting to and from work, vacations and business travel.

The 1973 OIL CRISIS was triggered by the OPEC oil embargo in response to the US decision to re-supply the Israeli military during the Yom Kippur war. The 1973 oil crisis started in October 1973, when the members of Organization of Arab Petroleum Exporting Countries or the OAPEC proclaimed an oil embargo "in response to the U.S. decision to re-supply the Israeli military" during the Yom Kippur war. It started October 1973and lasted until March 1974. Arab oil producers had also linked the end of the embargo with successful U.S. efforts to create peace in the Middle East, which complicated the situation. By January 18, 1974, Secretary of State Henry Kissinger had negotiated an Israeli troop withdrawal from parts of the Sinai. The promise of a negotiated settlement between Israel and Syria was sufficient to convince Arab oil producers to lift the embargo in March 1974. By May, Israel agreed to withdraw from some parts of the Golan Heights.

> *The United States and the World had been held hostage to the demands of those who controlled oil and in many cases were our enemies. The average consumer finally became aware that there may be problems with oil in the future. This was warning number one. We didn't listen.*

The 1975 MOVIE THREE DAYS OF THE CONDOR starring Robert Redford was a great thriller, but the reason for the chase and conspiracy is not clear until the end when you see that it was all about oil, and the fact the government of the United States was making contingency plans to fight for oil if necessary. The movie was absolute fiction. The United States would never think of going to war over oil.

APRIL 18, 1977, President Jimmy Carter gave a televised speech on our looming energy problems and the necessity for energy conservation. His speech is included later in this book. Anyone concerned about high gasoline prices or the future of energy should read that speech word for word.

1979 OIL CRISIS in the United States occurred in the wake of the Iranian Revolution. Amid massive protests, the Shah of Iran, Mohammad Reza Pahlavi, fled his country in early 1979 and the Ayatollah Khomeini soon became the new leader of Iran. Protests severely disrupted the Iranian oil sector, with production being greatly curtailed and exports suspended. When oil exports were later resumed under the new regime, they were inconsistent and at a lower volume, which pushed up prices. Saudi Arabia and other OPEC nations, under the presidency of Dr. Mana Alotaiba increased production to offset the decline, and the overall loss in production was about 4 percent. However, a widespread panic resulted, added to the decision of U.S. President Jimmy Carter ordering cessation of Iranian imports to the U.S., driving the price far higher than would be expected under normal circumstances. In 1980, following the Iraqi invasion of Iran, oil production in Iran nearly stopped, and Iraq's oil production was severely cut as well. After 1980, oil prices began a 20-year decline down to a 60 percent price drop in the 1990s. Oil exporters such as Mexico, Nigeria, and Venezuela expanded production, and the USSR for the first time became a world producer of oil. The US accelerated production of oil from Prudhoe Bay and Europe expanded oil production in the North Sea.

JULY 15, 1979, President Jimmy Carter delivered a televised speech. The speech was titled "Crisis of Confidence" and can be viewed on the internet. One small section of that speech describes his concern about the looming energy crisis.

"Ten days ago I had planned to speak to you again about a very important subject – energy. For the fifth time I would have described the urgency of the problem and laid out a series of legislative recommendations to the Congress. But as I was preparing to speak, I began to ask myself the same question that I now know has been troubling many of you. Why have we not been able to get together as a nation to resolve our serious energy problem?"

MARCH 24, 1989 the Exxon Valdez, an oil tanker bound for Long Beach, California, struck Prince William Sound's Bligh Reef spilling as much as 750,000 barrels of crude oil. It is considered to be one of the most devastating human-caused environmental disasters. Prince William Sound's remote location, accessible only by helicopter, plane, and boat, made government and industry response efforts difficult and severely taxed existing plans for response. The region is a habitat for salmon, sea otters, seals and seabirds. The oil eventually covered 1,300 miles of coastline, and 11,000 square miles of ocean. Again, the world became aware of what it takes to provide fuel for the family car.

AUGUST 8, 1990 Iraq under the leadership of Saddam Hussein annexed Kuwait for the purpose of obtaining oil money that he needed for his conquests. The United States realized the consequences if Saddam Hussein controlled such large oil reserves, which left no alternative to getting involved. After diplomatic maneuvering, February 23, 1991 a ground war begins with the objective of freeing Kuwait from the takeover by Saddam Hussein.

JANUARY 1991 fires were started by Iraqi in the Kuwait oil fields. The Kuwaiti oil fires were caused by Iraqi military forces setting fire to 700 oil wells as part of a scorched earth policy while retreating from Kuwait in 1991. Around 6 million barrels of oil were lost each day until the last oil well fire was extinguished in November 1991.

> *Why did Saddam Hussein invade Kuwait? Why did the United States commit to freeing Kuwait? Why did Saddam Hussein want the oil wells set on fire? Answer to all these questions is simply: OIL*

APRIL 20, 2010 was the day that awoke us to the fact that drilling for oil was a very important part of our lives. It was the day an explosion aboard the Deepwater Horizon oil platform in the Gulf of Mexico caused an environmental disaster which changed our thinking about oil, and alerted us to the fact that we were taking enormous risks to obtain oil. Again, there were ominous warnings that a primary safety device on offshore drilling rigs could fail at a critical time resulting in an environmental disaster. Specifically, the U.S. Minerals Management Service funded a study that was conducted by West Engineering Services titled Shear Ram Capabilities Study and dated September 2004. The following paragraphs are taken from that report:

WEST ENGINEERING SERVICES was commissioned by the United States Minerals Management Service (MMS) to perform a Shear Ram Capabilities Study. The main goal of the study was to answer the question "Can a rig's blowout preventer (BOP) be expected to prevent a blowout?" Shear rams may be a drilling operation's last line of defense for safety and environmental protection. The study concluded that they probably would not prevent a blowout.

> *Why did the U.S. Minerals Management Service fail to take appropriate action? Was it because they were incompetent, or was it because they were under pressure to keep the oil rigs drilling? Answer: drill baby drill drill drill.*

IMMEDIATELY AFTER the Deepwater Horizon oil platform exploded in the Gulf of Mexico new drilling was suspended indefinitely.

MARCH 23, 2011, less than a year later, offshore drilling permits were issued in spite of the fact that the cause of the Deepwater Horizon disaster had not been fully determined and no corrective action taken on the rigs currently operating in the Gulf of Mexico. The same Blowout Preventers (BOP's) that failed when the Deepwater Horizon exploded are still being used on new rigs in spite of the fact that they did not do their intended job.

> *Why did the U.S. Bureau of Ocean Energy Management, Regulation and Enforcement (BOEMRE) formerly known as the Minerals Management Service (MMS) issue Offshore Drilling Permits knowing that the Blow Out Preventers responsible for the Deepwater Horizon disaster had not been redesigned and proven to be capable of preventing another similar disaster? Same answer: Drill baby drill drill drill.*

These events show that the economy of United States is dependent on oil and government agencies are not willing to interrupt our essential flow of oil.

The beginning of the 21st century produced many "Doom and Gloom" books, headlines and documentaries about a looming energy crises and our dependence on foreign oil. The author decided to investigate to determine how bad the looming energy crisis really is and what can be done to prevent economic disaster. There is no simple answer. The best and simplest answer would be;

Act now, and act like we are at war because we are in an energy war and we are losing.

The looming energy crisis cannot be blamed on any single group of people or event. It is the combination of several things converging in the 3rd millinium.

1. Population Growth
2. Technological Advancements
3. Deterioration of Leadership
4. Human Greed and Selfishness

SECTION ONE
INTRODUCTION

Eventually the world is going to have an energy crisis. There are very few government or industrial leaders that dispute this fact. The only dispute is whether eventually means the energy crisis is already upon us, or that it will occur in a year or a hundred years. There is no one person or group that will resolve the energy problem. It will take the efforts and sacrifice of all.

EFFORTS OF ALL

<u>Government leaders</u> have access to information proving that the world is approaching depletion of its fossil energy reserves. They are the ones that must understand, and convince their constituents that dependence on fossil fuel is coming to an end and "drill baby drill drill drill" is not the answer to the inevitable problem of depleting fossil fuels. Politicians must convince voters we must make unpopular decisions, such as conservation and rationing in order to prevent a future energy crisis that will have drastic consequences on the world economy. Government leaders, particularly in the United States are taking actions on the depleting fossil reserves that are too little and too late to prevent suffering and starvation. They must understand the approaching energy problem and how it will affect the economy. This book may help.

<u>Geologists</u> are the technical people who can figure out how much fossil fuel reserves we have left in the earth, and how economically practical it will be to extract those fuels. However, geologists are employed by the corporations that will benefit from our depleting fossil reserves and the resulting increases in fuel prices which will benefit them. Because of their loyalty to the big oil corporations, their reports and statistics may not be entirely accurate. Geologists also have the responsibility of determining if and where geothermal energy can be economically produced well beyond the present availability.

Agronomists are the technical people who can figure out whether growing ethanol fuel or algae fuel will be practical in the future when increases in world population necessitates growing food rather than growing fuel. When oil is depleted, bio-fuels may be the only practical mobility fuels for certain conveyances such as aircraft and ships.

Engineers will provide us with products that will allow us to cope with the limited amount of fossil fuel energy still available and will provide us with renewable energy from the sun, wind and ocean currents. Engineers will design transportation systems that will operate on available electrical energy.

Architects are designing new "centers" that will be accessible by energy efficient mass transportation. Heating and air conditioning will be more efficient by using internal atriums with less surface contact to the outside air to reduce loss of energy. The centers for living, shopping, health, education, industry and many more will be connected by rapid mass transportation. These centers will not have inefficient windows with beautiful views of the surrounding landscape, but, rather will have views of internal atriums. The wealthy that have beautiful views of the surrounding landscape will be considered wasteful squanderers of our depleting energy resources.

Biologists are working on renewable sources of mobility fuels for aircraft grown from algae. These have proved to be feasible when, November 2011, the first commercial flight using algae biofuel flew from Houston, Texas to Chicago O'Hare airport.

Physicists are presently working on fourth generation safe nuclear reactors that could provide us with enough electricity for thousands of years.

Financiers will bring together resources necessary to develop and implement costly nuclear and geothermal projects. Considering the reputation that financial institutions received during the housing melt down, this will prove to be a difficult task.

Every individual must understand that we are facing an energy crisis and that squandering our fossil fuels by driving around in a large vehicle is not the symbol of success but rather is the symbol of ignorance and selfishness.

The idea for this book came on April 20, 2010. It is the day the Deep Water Horizon drilling rig in the Gulf of Mexico exploded, resulting in the greatest environmental disaster in United States history. That disaster pointed to the fact our government and industrial leaders were making energy decisions based on panic and greed without considering the future of this planet.

Research of books and internet articles on the subject of depleting natural resources showed many gloom and doom books and articles. Internet research revealed reams of data from reliable sources as well as some questionable sources. Studying those many reams of data available on the internet made it apparent something had to be done to digest data and present it in a form that could be understood by the average student, politician, and industrial leader. That is the intention of this book.

The average person is not interested in the future environment until it directly affects his everyday life. The future for the average person is retirement for which he will plan to have enough assets for food, a home, medical care and leisure. The future 100 years from now is of little concern to him.

Many people have immediate personal problems to occupy their time and energy without being concerned with the future, regardless how distant it may be. This poses a question as to who the people are that are concerned about the future? Is it the average person? Is it the Congress of the United States Government? Is it the United States Environmental Protection Agency? Is it private groups such as Greenpeace, Friends of the Earth, The Sierra Club or any of the activist groups which make meager headway against the big corporations whose objectives are financial gain rather than concerns about the future environment, particularly the very distant future environment? Future environment and very distant future environment are very broad terms if we attempt to define them in terms of years. This book will make an attempt to broadly define those terms.

The 20th century witnessed technology changes that are so complex and difficult to understand only a few intellectuals can fully understand the long term effect they will impose on the world. One of the great technological accomplishments is the internet, or what many refer to as the information highway. It is impossible to estimate the enormous quantity of information available on the internet

This book is intended to provide concerned citizens, educators, business leaders and politicians with information obtained from numerous sources so they may have a better understanding of the seriousness of our looming energy crisis.

This book discusses subjects that our political leaders avoid because of the potential negative effects on their political careers. One of the basic human rights that every person has is to have children and raise a family. Limiting that right in a democratic society such as the United States is political suicide. We are facing eventual depletion of energy derived from fossil fuels and are already seeing the economic consequences of importing oil from countries that are our enemies. No matter how critical the energy situation becomes, a politician who recommends energy rationing or population controls would have a very limited career. Increasing taxes has always been a subject that politicians fear to discuss no matter how our economic future deteriorates.

SOME OF THE QUESTIONS THAT THIS BOOK WILL ATTEMPT TO ANSWER INCLUDE:

1. Is population one of the main causes of our looming energy crisis?
2. Why does the United States have low gasoline prices?
3. Is gasoline rationing an equitable way of reducing conumption of mobility fuels?
4. What will be the economic ramifications if we do not reduce energy consumption, particularly gasoline?
5. How much will sustainable energy sources, such as wind power and biofuels alleviate oil depletion?
6. Will nuclear energy evolve as a safe and environmentally friendly source for electrical energy?
7. Was it a mistake to allow cheap energy to drive our economy knowing that cheap energy could not last forever?
8. Will our political leaders have enough courage to take action that may be painful to their voting constituents?

This book should be compulsory reading for all government leaders and politicians

There is a lot of false and misleading information on the internet and in books. While writing this book, every effort was made to provide accurate information on the subjects addressed in this book.

Here is an article posted on the internet 02/13/12 - At a conference in Philadelphia last October, a Wharton professor noted that one of the country's biggest economic problems is a tsunami of misinformation. You can't have a rational debate when facts are so easily supplanted by overreaching statements, broad generalizations, and misconceptions. And if you can't have a rational debate, how does anything important get done? As author William Feather once advised, "Beware of the person who can't be bothered by details." There seems to be no shortage of those people lately.

SECTION TWO
POPULATION

Propagation of the human race is one of the most controversial subjects of politics, religion and human rights. There have been religious, scientific and layman writings on the subject. Some of those writings have taken the position that the earth can support a limited number of human beings without starvation and suffering. The subject of controlling population is controversial to the extent that democratically elected politicians have avoided it. It is considered the third rail of politics because having children and raising a family are considered a basic right, privilege and obligation of human beings.

Population is expressed as a quantitative number, but it does not include a number that expresses the quality of life. It is simple to understand that as population increases, quality of life must decline if there are finite resources to support that population.

It was not until the 20th century that leaders of overpopulated nations have attempted to limit population growth. China has achieved limited success and India has achieved very little success.

Since the beginning of recorded civilization, population has been increasing at an alarming rate with the exception of some natural occurring disasters prior to the 16th centuty.

The planet Earth has 57 million square miles of land surface of which only 10% is arable. It has finite amounts of minerals, some of which are abundant and some of which are being depleted at an alarming rate. Once depleted, those minerals will never be replaced. As of the end of the 21st century, there will be 8 billion people consuming those minerals, with starvation and economic suffering becoming more common throughout the world, including the wealthy developed nations.

Consumption of minerals, particularly fossil fuels by an ever increasing population is the subject of this book. A good start is the Malthusian Theory of Population.

MALTHUSIAN THEORY OF POPULATION

Malthusian theory of population was originally foreseen to be a forced return to subsistence-level conditions once population growth had outpaced agricultural production. The term is also commonly used in discussions of oil depletion. His theory suggested that growing population rates would contribute to a rising supply of labor that would inevitably lower wages. In essence, Malthus feared that continued population growth would lend itself to poverty.

His book "An Essay on the Principle of Population" was first published anonymously in 1798. The author was soon identified as The Reverend Thomas Robert Malthus. While it was not the first book on population, it has been acknowledged as the most influential work of its era.

The Malthusian Theory was further confirmed in a front page Wall Street Journal article in March 2008 pointed out various limited resources which may soon limit human population growth because of a widespread belief in the importance of prosperity for every individual and the rising consumption trends of large developing nations such as China and India.

These theories reared their ugly head when the national debt of the United States threatened to lower its credit rating for the first time in history. August 1, 2011 was the deadline for increasing the debt ceiling in order to avoid default on many obligations primarily that of debt service on outstanding treasury notes.

July, 2011 witnessed many political debates on how to handle the looming debt crisis. Causes of the debt crisis were known to be exorbitant spending on social entitlement programs and the high costs of Social Security and Medicare, all of which were in some way connected to population increases that were not considered when the original programs were designed and put into law.

Many proposals were considered, none of which addressed the underlying cause of the world economic crisis which is our increasing population combined with depleting oil reserves.

WORLD POPULATION

World population and world energy consumption have been increasing at rates this fragile planet cannot sustain indefinitely without intolerable suffering and starvation.

World population has been increasing steadily since about 1500. This is true even though there have been many wars that killed millions of people. Combined with naturally occurring disasters such as earthquakes and floods, these events did little to stem the increase in world population.

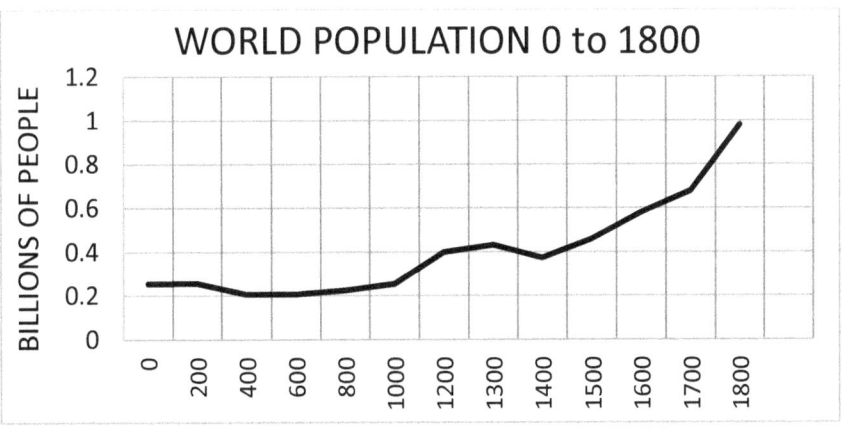

UN DEPARTMENT OF ECONOMIC AND SOCIAL AFFAIRS

Reviewing the population graph for the years 0 to 1800, there is a decline from about 200 to 1000. No organization has offered an understanding of why this occurred.

There is another decline between 1300 and 1400. History provides an explanation for the decline in population in the year 1340 which was caused by the Black Death or Plague. The Black Death was one of the most devastating pandemics in human history, peaking in Europe between 1348 and 1350. The Black Death is estimated to have killed 30% – 60% of Europe's population reducing the world's population from an estimated 450 million to between 350 and 375 million in 1400. It took 150 years for Europe's population to recover.

Modern medicine of the 21st century makes it highly unlikely that catastrophic diseases such as the Plague will reoccur in the future.

After the 15th century world population had a steady increase, with man-made disasters such as war, natural disasters, floods, earthquakes, tsunamis, and tornadoes having little effect on the steady increase in world population. Later paragraphs list some of the more tragic Man Made Disasters and Natural Disasters.

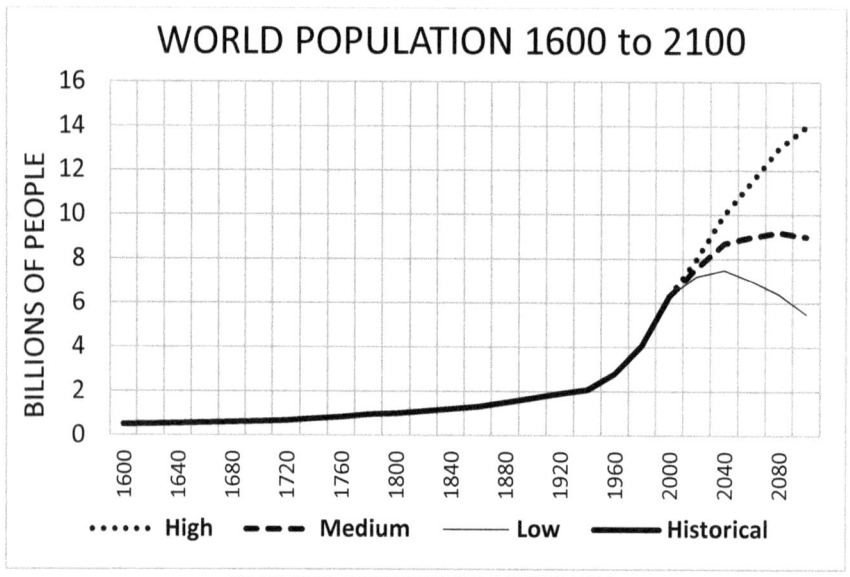

US BUREAU OF THE CENSUS 1600 TO 2000
UN DEPARTMENT OF ECONOMIC AND SOCIAL AFFAIRS 2000 TO 2080

The United Nations Department of Economic Affairs is the only agency that has attempted to estimate population 70 years into the future. Difficulty in making such predictions is shown by the estimates for the year 2080 which have a high of 14 billion people, and a low of less than 6 billion, the latter reflecting a decline in world population which is highly speculative.

One of the major factors in the exponential increase in population since the start of the industrial revolution is the increased life expectancy resulting from improvements in medical care. Further reasons for the increased rate of population growth are discussed later in this section.

WORLD'S MOST POPULOUS COUNTRIES AND PREDICTED POPULATION GROWTH MILLIONS OF PEOPLE

	2000	2050	% growth
India	1,015,000	1,620,000	60
China	1,262,000	1,470,000	16
USA	275,500	404,000	46
Indonesia	225,000	338,000	50
Nigeria	123,000	303,000	146
Pakistan	141,500	267,800	89
Brazil	172,800	206,700	20
Bangladesh	129,000	205,000	59
Ethiopia	64,100	188,000	193
Congo	51,900	182,000	250
Philippines	81,100	154,000	89
Mexico	100,000	153,000	53
Vietnam	78,700	119,000	51
Russia	146,000	118,200	-19
Egypt	68,300	113,000	65
Japan	1265,00	101,200	-20
Iran	65,600	100,200	53
Saudi Arabia	22,000	91,100	313
Tanzania	35,300	88,300	150
Turkey	65,600	86,500	32

The above tabulation shows that China's one-child policy, in which, with various exceptions, having more than one child is discouraged, is shown to be effective when compared to other developing countries such as India. Unauthorized births are punished by fines, although there have also been allegations of illegal forced abortions and forced sterilization. The Chinese government introduced the policy in 1978 to alleviate the social and environmental problems of China. According to government officials, the policy has helped prevent 400 million births. The success of the policy has been questioned, and reduction in fertility has also been attributed to the modernization of China. The policy is controversial both within and outside of China because of the issues it raises, the manner in which the policy has been implemented and because of concerns about negative economic and social consequences.

Family planning in India has had little effect on its growing population. Newly married couples are asked to wait two years before becoming pregnant, and the government will thank you. It also will pay 5,000 rupees, or about $106, if the couple delays having children. Waiting also would allow India more time to curb a rapidly growing population that threatens to turn its demography from a prized asset into a crippling burden. With almost 1.2 billion people, India is disproportionately young; roughly half the population is younger than 25. This "demographic dividend" is one reason some economists predict that India could surpass China in economic growth rates within five years. India will have a young, vast work force while a rapidly aging China will face the burden of supporting an older population.

India is expected to surpass China as the most populous country by the year 2030.

The above tabulation shows that the greatest population growth rates are occurring in undeveloped African countries such as Congo, Ethiopia, Nigeria, and Tanzania.

An exception to this is Saudi Arabia's population which as of the April 2010 Census was 27,136,977 consisted of 18,707,576 Saudi nationals and 8,429,401 non-nationals. Saudi Arabia's population is characterized by rapid growth and a large cohort of youths. Population Growth Rate is 1.536% and the Total Fertility Rate is 2.35 children born per woman.

WORLD POPULATION AND GROSS DOMESTIC PRODUCTION PER CAPITA 2009

	2009 Population	1999 GDP per Capita US $	2009 GDP per Capita US $	% GDP per Capita Growth
Liechtenstein	34,761	82,020	134,392	63
Luxembourg	491,775	49,219	105,044	113
Bermuda	67,837	53,581	88,747	66
Norway	4,660,000	35,660	79,089	122
Qatar	833,285	21,020	69,754	232
Kuwait	2,691,000	14,296	54,260	279
Ireland	4,203,000	25,643	51,049	99
United States	307,212,000	33,300	46,000	38
Singapore	4,657,542	20,868	36,537	74
China	1,340,000,000	865	3,744	333
India	1,166,000,000	451	1,192	164

The United States which is the wealthiest nation in the world has one of the lowest GDP (Gross Domestic Production) per capita growth rates. The explanation can be complacency, politics, obligation to protect the world, greed by a few causing the 2008 recession, or depletion of some of our resources.

The above figures are average GDP per capita. They are not an indication of the distribution of personal wealth. Personal wealth should not be confused with income which is the amount a person or family earns in a specific period of time, normally a year. Personal wealth is a person or families assets minus liabilities which have accumulated during their lifetime.

Liechtenstein, Qatar, Luxembourg, and Bermuda, which in addition to being small countries have unique reasons for their wealth other than oil.

In the middle of the GPD per capita tabulation are the oil producing nations which are wealthy because of the well-known theory of supply and demand.

Kuwait's economy is free and open by its nature and is one of the most prosperous in the world and is mainly backed by its crude oil production. Kuwait is known to have the largest reserves of crude oil. The Economy of Kuwait is mainly based on the petroleum industry. The country is one of the richest countries in the world. Since the Gulf War, it has gone through a fast reconstruction process.

Qatar has the world's largest per capita production and proven reserves of both oil and natural gas. The main drivers for this rapid growth are attributed to ongoing increases in production and exports of liquefied natural gas, oil, petrochemicals and related industries after the United Arab Emirates

At the low end of the GPD per capita tabulation are the developing nations with large populations such as China and India. Their low GDP per capita is attributed to the low ratio of natural resources to population. The resource that both of these nations have in common is their population which is attributed to their recent growth in GDP per capita.

China's GDP per capita increased between 1999 and 2009, by 333% while the US GDP per capita increased by only 38%.
Predictions by credible sources have China's economy overtaking the US sometime between 2019 and 2030. China remains far behind the developed world in per capita terms, and because there is plenty of catch-up left to accomplish, there's plenty of room for rapid growth, and China has the population to accomplish this.

India's GDP per capita increased between 1999 and 2009, by 164% while the US GDP per capita increased by only 38%. However, this increase was less than half that of China for the same period. Part of the answer is that India's greater population growth diluted the benefits of their larger national GDP.

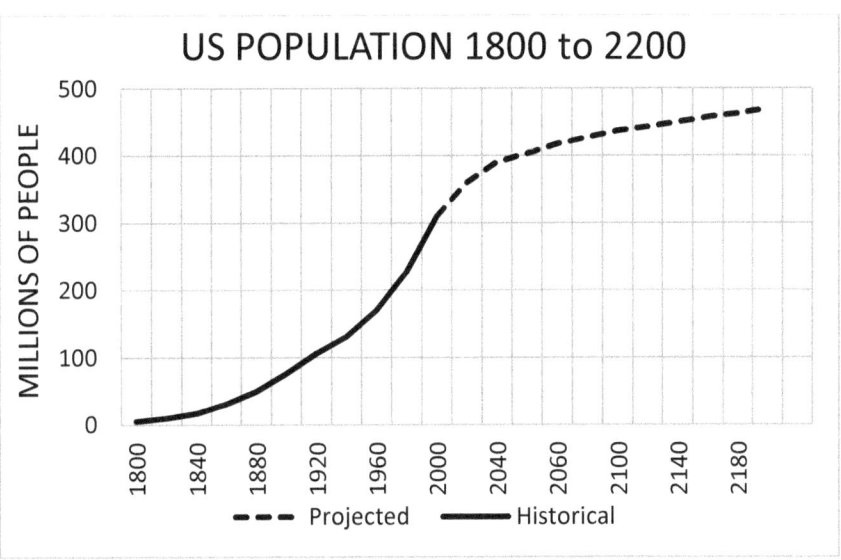

US BUREAU OF THE CENSUS 1800 TO 2000
UN DEPARTMENT OF ECONOMIC AND SOCIAL AFFAIRS 2000 TO 2200

The United States is a relatively new country, less than 250 years old, with population growth effected by annexation of territories and liberal immigration policies. Estimates for the years 2010 to 2200 depend upon what population control measures and immigration policies are adopted by the political leaders of the United States.

The United States Social Security System is one of the factors that may be controlling population growth. Necessity for families to have large numbers of children to support them in time of need and old age has been replaced by the Social Security System and modern farm machinery has replaced the need for families to have large numbers of children to run the farm.

The United States is one of the most liberal countries in the world, which unknowingly provides tax and welfare incentives for large families. The above graph titled "US Population 1800 to 2200" shows a trend towards a peak in population during the 22nd century. If our political leaders can make this a reality, we may be able to concentrate our efforts to save the rest of our planet.

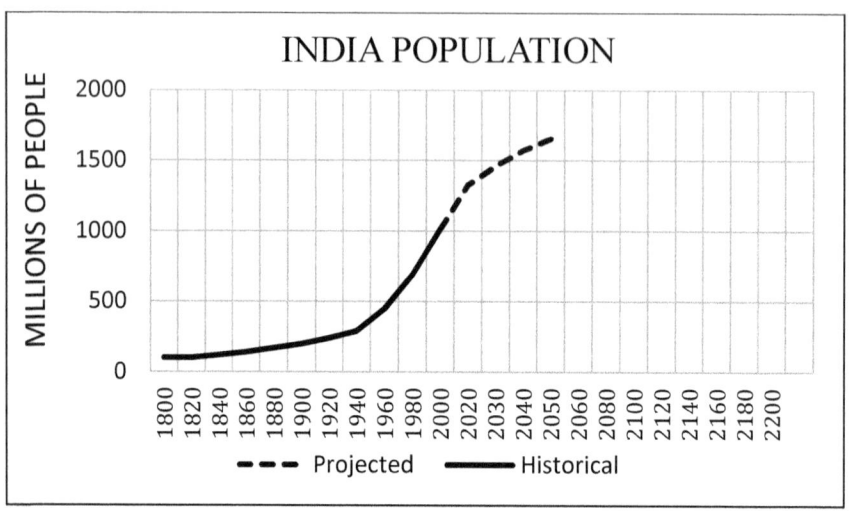

THESE POPULATION FIGURES DO NOT INCLUDE EAST AND WEST PAKISTAN PRIOR TO PARTITIONING IN 1947. PPROJECTIONS BEYOND 2050 ARE NOT AVAILABLE DUE TO THE VOLATILITY OF INDIA POPULATION GROWTH.

India has a universal health care system run by the local governments. Government hospitals provide treatment at taxpayer expense. Most essential drugs are offered free of charge in these hospitals.

India's population growth rate of 1.38% annually is alarming the Indian government. India is projected to overtake China as the world's most populous nation by 2030. India's population growth has raised concerns that it would lead to widespread unemployment and political instability.

INDIA POPULATION CONTROL

India is addressing birth control by providing couples with a monetary incentive if they decide to postpone plans for a child for at least two years after marriage. The government is offering $106, a significant sum in India's rural areas, if they agree to its rules. The program was first launched in Satara, Maharashtra, a state in Western India, with already more than 2000 couples reported to have enrolled for the program.

Although India has population control incentives, it does not have mandatory population controls that can achieve the results of the Chinese. Reason for the ineffective population control in India may be the result of the type of government. India has a multiparty political system with elections. Voters in India view raising a family as a basic human right that cannot be taken away by the government. A political party that proposes mandatory population controls would be very controversial. Further reason for the inability to control population is ineffectiveness of the government, particularly in the area of education where groundwork for population control must be established.

CHINA POPULATION CONTROL

Population control was introduced in China in 1978, referred to as the One-Child Policy. Success of the policy is yet to be determined. However, figures published by the United Nations show that from 1950 to 2000, population increased from 554 million to 1,275 million, an increase of 720 million. However, projections to 2050 show a population of 1,395 million people, an increase of 120 million As a comparison, India which does not have an effective population control policy, had almost a 200% increase in population from 357 million to 1,017 million during the period 1950 to 2000. Population of India is projected to increase another 50% from 2000 to 2050 to 1,531 million and will pass China to become the most populated country in the world by 2050.

Depending upon how the statistics are reviewed, it can be concluded that the population control measures in China had a significant effect on the rate of growth of the population of China. However, the population of China continues to grow at an alarming rate. There were side effects, some are against the basic principles of some nations, and some can be considered an inconvenience, one of which is the increase in the ratio of males to females.

India which has no mandatory population control measures continues to grow at an alarming rate and will be the most populous nation by the year 2050.

Parents choose to have smaller families when they do not have to rely on their children to work on the family farm or business or to take care of them in their old age.

SOCIAL SECURITY FOR UNDEVELOPED NATIONS

Governments must face the fact that one of the primary reasons families have children is to provide for their security in later years and in event of illness. If governments provided a dependable form of social security, families would no longer depend on their children to support them in time of need and old age.

Governments of developing nations could intervene in population growth by providing an <u>incentive</u> type of social security. For purposes of this proposal families would be discouraged from having more than one child.

1. An only child in the family and the child's parents are expected to pay a small percentage of their earnings into a fund to be distributed by the government to the elderly and needy.
2. Two or more children in a family and the children's parents are expected to pay a higher percentage of their earnings into the fund.
3. Persons of working age who do not have employment and cannot make payments into the fund will be expected to work for the agency to build retirement facilities, or work in a retirement facility until they are able to find other employment.
4. Married couples who do not have children and single adults will pay the least amount into the fund. Upon retirement they will be eligible for retirement at a state run facility or could elect to have a payment applied to a private facility.

MAN MADE DISASTERS

In 2011 the Japanese city of Fukushima experienced a devastating tsunami. Was it a natural disaster or was it a man-made disaster? It was a manmade disaster because man decided to build a nuclear power plant on a seismic fault. What choice did the planners of the nuclear facility have? Not much. Japan has very little territory and many people. Why do we build cities on seismic faults? Because when the cities were founded there was no indication they were on a seismic fault. The entire question of manmade disasters from seismic faults, earthquakes, floods can be addressed by a single word – population.

Manmade disasters such as conventional war, revolutions, biological war, and nuclear power plant radiation accidents will continue to occur. Technology is a two edged sword. On one hand, technology will give us means to gather intelligence to prevent war. On the other hand, technology will give rogue nations greater ability to inflict harm on peaceful nations. Extent of population reduction by man-made disasters is impossible to predict.

A thought that occurs in people's minds is a nuclear holocaust by a rogue nation or terrorist organization. One weapon that can be used by free nations against such attacks is modern technological communications to educate freedom loving citizens so that they will stop "the bad guys" whether they are the dictatorial leaders of suppressed countries or leaders of small terrorist networks.

Modern technology is responsible for enormous consumption of energy resulting in global warming. Fortunately, scientists recognize the danger of global warming on our environment and are slowly taking action to preserve our planet. Reluctance of industrial leaders to take action could result in an environmental disaster that may be irreversible.

Comparing the devastation caused by all man-made disasters, war has been the greatest cause of loss of human life, with World War II, from 1939 to 1945 claiming 60-70 million lives, the most of any war.

There have been many wars, too numerous to list which have accounted for millions of human deaths. Wars are man-made disasters which could have been prevented, and claimed a horrible amount of deaths. In spite of wars, population growth has continued.

Famine must be considered a man-made disaster because man could have prevented famine by controlling population and resources. Famine has caused the second greatest loss of human life in the past, the China Famine of 1907 claimed 24 million lives. Unfortunately, population increases and poverty will continue to take an unnecessary toll of human lives in underdeveloped and overpopulated nations.

There is a man-made disaster looming in the future. The world has been told, but is not willing to, or able to resolve it. Ask Al Gore about it.

NATURAL DISASTERS

Natural disasters include disease, famines, avalanches, blizzards, cyclones, earthquakes, floods, heat waves, tornadoes, tsunamis, volcanoes, and wild fires.

Historically, only deaths from disease have eclipsed the number of deaths due to war. With the advent of modern medicine, deaths from disease must be taken out of the category of being natural disasters and classified as man-made disasters, because man has the ability to prevent them.

Disease has caused the greatest loss of human life in the past. Modern medicine has eliminated or has the ability to eliminate all of these diseases.

Smallpox	1900-1980	300 million
Malaria	1900-2000	80-250 million
Black Death	1300-1720	100 million
Spanish Flu	1918-1920	50-100 million
Tuberculosis	1900-2000	40-100 million

Floods have caused the third greatest loss of human life, with the China Floods of 1931 taking 2.5-3.7 million lives. Both the Yangtze and the Yellow Rivers are being tamed by the Yellow River Conservancy and the Three Forks Dam respectively. If these projects are successful, less devastation can be expected from floods in China.

Another conclusion is that world population continues to grow in spite of losses of hundreds of millions of lives caused by both natural disasters and man-made disasters.

WHY DOES POPULATION INCREASE?

Increases in population resulting from immigration are very simple to calculate for any individual country. It is a simple matter of addition. Increases in world population are not directly affected by immigration from one nation to another. The net result is zero. Therefore, let us put immigration aside and talk about total fertility rate, which is somewhat more difficult to comprehend.

Simply stated, if there were no mortality in the female population until the end of the childbearing years, then the replacement level that would maintain world population at a steady level would be a total fertility rate (TFR) slightly higher than 2.0, meaning that every woman would have slightly higher than 2.0 children.

However, the replacement level is also affected by mortality, especially childhood mortality. The replacement fertility rate is roughly 2.1 births per woman for most industrialized countries but ranges from 2.5 to 3.3 in developing countries because of higher mortality rates. Taken globally, the total fertility rate at replacement is 2.33 children per woman. At this rate, global population growth would trend towards zero.

A population that maintained a TFR of 3.8 over an extended period of time without a correspondingly high death rate would increase rapidly, whereas a population that maintained a TFR of less than 2.0 over a long time would decline.

Some general reasons that total fertility rates vary from one nation to another are:

1. Underdeveloped nations lack sex education for family planning and lack contraceptive devices to achieve birth control, thus resulting in unplanned and unwanted children.
2. Underdeveloped nations have TFR as high as 3.8.
3. Developing nations such as China and India do not have a Social Security system similar to that available in the United States

4. Developing nations have a TFR of 2.3 or more.
5. Industrialized nations provide education in sex and family planning which attribute to a relatively low TFR. In industrialized nations, business careers take precedent over raising a family, resulting in a TFR of less than 2.0. An example of this is the country/state of Singapore which had a TFR of only 1.29 in 2007.

The above are general reasons for population increases. Each nation has its own reasons for population increases. For example, the United States is a new nation with population growth effected by:

1. Annexation of territories
2. Liberal immigration
3. Welfare incentives to unwed mothers
4. Tax incentives for families with children
5. Modern technology and medicine make rearing a family less of a burden

Political leaders must decide if it is in the best interests of the nation to increase population by allowing liberal immigration of unskilled labor which burdens the already expensive entitlement programs. An immigration policy that puts an incentive on the emigrating (outflow of people) nation to reduce family size is a political issue that politicians are unwilling to get involved with.

All creatures are created with the natural sexual instinct to perpetuate life. This is true for human beings as well as all animal life. A very good example of this animal instinct is shown in the award winning movie "March of the Penguins".

A little humor can be shown about sexual instinct by looking at the electrical power failure that occurred on August 14, 2003 in the Northeast United States. Precisely nine months after that power failure, there was a surge in births in the regions where the power failure occurred. This needs little explanation.

Prior to the Social Security Act of 1935, families depended on their children to provide for them in their later years, or in the event of a physical impairment that deprives them of an income for essential needs and services. The need to have large families as a means of "Social Security" was no longer necessary after the

Social Security Act was introduced in the United States. Statistics show this to be true with a gradual decline of the size of families after 1935.

Large families were still needed to work the farms prior to the introduction of modern farm machinery. After the 1960's when corporations took over the farming industry, large families were no longer necessary to work the farms.

WHERE DOES POPULATION INCREASE?

Business opportunities, resources, employment and transportation are the main reason that people settle in a particular location. Transportation by water continues to be the most economical method to move industrial products. Maritime transportation hubs require large navigable harbors surrounded by cities built on low lying terrain. These cities are just a few feet above sea level making them susceptible to several forms of disaster. Hurricanes form over open oceans sometimes traveling towards land where they have devastated coastal areas. Tsunamis caused by earthquakes under the oceans have devastated low lying coastal areas, the most publicized of which is the Fukushima, Japan disaster of 2011 which caused a major nuclear meltdown. Hurricanes and Tsunamis, both of which are natural disasters, attack low lying coastal areas where populations are most dense. Low lying coastal areas are now susceptible to man-made disaster resulting from global warming. As the temperature of the earth rises, glaciers will melt causing ocean levels to rise, eventually inundating coastal cities. Environmentalists claim that this is already occurring. As the coastal cities become inundated, inhabitants will seek higher ground, causing inland cities with limited resources to accommodate more people resulting in overcrowding and economic burden.

Global warming will be a man-made disaster that could have been prevented. It will result in population reduction accompanied with starvation and suffering.

OPTIMUM POPULATION

How many people can this planet accommodate without starvation, disease and suffering? What is the optimum population for the world? What is the optimum population for the United States, China, India or Luxembourg? Those are the questions that have increasingly been the subject of many books and public debates for the past century. The consensus of opinion is that the planet is becoming overpopulated and we are starting to see the effects of overpopulation on the world environment and depletion of natural resources.

It is impossible to put any kind of a figure on the optimum population for the world or any single nation of the world. Each nation has its own unique problems. Some nations recognize the problem of overpopulation but have not been able to correct the problem.

The answer to the question of optimum population can be better understood by asking these questions. What kind of world would we live in today if 100 years ago world population started to level off at two billion people? At one billion people? Would there be enough food, water, and natural resources for the entire world? If the answer is yes, how long would food, water and natural resources last?

We do not know and will never know the answer to optimum population. What we do know is that less population will be less of a burden on our food, water and natural resources. An answer may be to say stop when there is not enough food, water or energy to support all the people of the world without hunger and suffering.

CONCLUSION

World population will continue to grow well into the 23rd century in spite of efforts by some governments to control birth rates. Industrialized nations are starting to show a decline in population, possibly because young couples are pursuing careers in business.

However, these declines are a small fraction of the population increases from underdeveloped nations. Combining these population increases with increasing energy consumption per capita points to consumption of our energy reserves faster than new reserves are being discovered.

In order to prevent starvation, disease and suffering, world political and industrial leaders must control population, energy consumption, and depletion of our limited natural resources.

The alternative to population control will be revolutions, war, famine and anarchy.

> *At this point, let us stop and say a prayer that population will not be reduced by a nuclear holocaust.*

SECTION THREE
LEADERSHIP

There are three types of influential people that affect the course of the world and of all the individuals in the world. They are dictators, politicians and true leaders.

Dictators have been around since the beginning of recorded history leaving their scars on the world. They are narcissists who care only about themselves and believe that they can do no wrong. Their objective is to maintain power by suppressing the people that they control. The last thing on their minds is to make the world a better place to live, certainly not two hundred years into the future.

Politicians are the product of free democracies that elect leaders by voting. They have the ability to influence people in order to get elected but in many cases are not the best leaders. Many are narcissists who believe that they can make decisions by their uncanny instinct, rather than devoting time and effort to study a problem in order to take the best course of action. They shoot from the hip. Fortunately, there are a few politicians that are true leaders – not enough.

True leaders have an objective, generally of good intentions and are willing to sacrifice of themselves and of their followers in order to achieve those objectives. True leaders have made their mark on the world and left it a better place to live. We name cities and streets after them. Unfortunately, there are not enough men with leadership potential that are willing to come forward and make the necessary sacrifices to lead.

Now at the beginning of the 21st century, we are facing an energy crisis, an economic crisis, and in several nations, a population crisis. Resolution of these crisis will require true leadership.

Corporations that have true leaders capable of managing complex structures successfully will succeed, grow and make a profit. They will provide innovative products and services that will make our nation grow. A corporation has many tools available to allow it to achieve its goals. A corporation can borrow money from banks or individuals by issuing stocks and bonds. If a corporation does not meet its financial obligations, it will cease to exist, or if it is too big to fail, it will be bailed out by the government. It can be creative with new products and services. Some corporations are monopolies in the sense of owning mineral rights or intellectual properties that it utilizes to the best extent possible. If necessary, a corporation can reduce its payroll by terminating its employees without any regard as to whether those employees will find employment elsewhere.

The United States government in comparison can be considered to be the largest corporation in the world. It can borrow money from individuals, both domestic and foreign, that must be paid back. Debt service on government borrowing is a burden on the economy similar to the burden of debt service on a corporation or an individual. It is doubtful as to how many of our politicians fully understand the effect of debt service on the economy or the individual taxpayers who must eventually pay back that debt.

The United States has mineral reserves that must be protected for future generations. Political leaders have the duty to protect those mineral resources but do not understand or do not want to understand the consequences of depleting mineral resources while the population continues to increase. It has many other obligations that are too complex for any individual to fully understand. The problem of providing a descent life for all its citizens while the population increases and its mineral resources become depleted may be too much for most leaders to understand. If our political leaders do not fully comprehend problems, how is it possible for them to resolve problems?

OBJECTIVES

Objectives should be the guiding force of our leadership. When we use the word objective, there must be a time element included such as short term, long term, future, or utopian. Unfortunately, most people, particularly our government and industrial leaders think of their objectives in the short term. Therefore, we must set a time element with a very broad range on objectives.

Short term objectives of our political leaders are to demonstrate that they are doing a good job, so that they will be reelected. Short term objectives of corporate leaders are the bottom line, knowing that the board of directors will give them a bonus and increase their salary if the price of the corporate stock increases in value. These appear to be very selfish objectives, but human beings are instinctively selfish.

Long term objectives of both government and industrial leaders are woefully lacking. The last thing on the minds of our government leaders is the world population explosion, depletion of arable lands, environmental pollution, last but not least is the simple fact that eventually the world oil supply will be depleted to the point that the food supply of the United States which is dependent upon oil will become inadequate. In the 21st century, corporations have introduced a new top level position named the CVO (Chief Visionary Officer). Responsibilities of the CVO are yet to be defined, particularly when it comes to how far into the future the corporation should be planning. Should they plan for a corporation with products to serve us 20 years from now, or 2000 years from now?

GOVERNMENT STRUCTURE

When the founding fathers of the United States wrote the constitution, they understood that government was so complex that the voting public was not capable of selecting the best leaders to run the government. This was kept in mind when the founding fathers wrote the Constitution.

Government structure is clearly defined in the constitution. However, the duties and responsibilities of the United States Government are neither well defined nor understood. With this in mind, the Preamble to the United States constitution, although very limited in nature, broadly states:

> *We the People of the United States, in Order to form a more perfect Union, establish Justice, insure domestic Tranquility, provide for the common defense, promote the general Welfare, and secure the Blessings of Liberty to ourselves and*

our Posterity, do ordain and establish this Constitution for the United States of America.

The constitution and its preamble had the best of intentions and made the United States of America the greatest nation in the world. Since our founding fathers wrote the Constitution, our nation has grown in population, technology and prosperity at a rate that could not possibly have been foreseen. At the time our Constitution was written, population of the United States was less than 4 million compared to almost 300 million at the beginning of the 21st century. Land area of the original 13 states was 865 thousand square miles compared to over 3.5 million square miles at the beginning of the 21st century. Advances in technology and prosperity could not possibly have been foreseen when our Constitution was written.

The founding fathers of the United States realized that we should not leave the election of the President of the United States up to people who were not qualified, by reason of education, information or prejudice. They feared a tyrant could manipulate public opinion and come to power. Alexander Hamilton wrote in the Federalist Papers:

It was equally desirable, that the immediate election should he made, by men most capable of analyzing the qualities adapted to the station, and acting under circumstances favorable to deliberation, and to a judicious combination of all the reasons and inducements which were proper to govern their choice. A small number of persons, selected by their fellow-citizens from the general mass, will be most likely to possess the information and discernment requisite to such complicated investigations. It was also peculiarly desirable to afford as little opportunity as possible to tumult and disorder. This evil was not least to be dreaded in the election of a magistrate, who was to have so important an agency in the administration of the government as the President of the United States. But the precautions which have been so happily concerted in the system under consideration, promise an effectual security against this mischief.

Hamilton and the other founders believed that the electors would be able to insure that only a qualified person becomes President. Credit must be given to our founding fathers for recognizing the fact that it will be difficult to elect the best persons to lead the nation. In this respect they were absolutely correct. Now in the 21st century, it will be extremely difficult to find a single person who will deny the fact that we made mistakes in the selection of our government leaders.

There is one modern day President of the United States, Harry S. Truman, who understood better than anyone how government functioned. Here are just a few of his quotes that sum up politics and the presidency. The most famous of which is "The buck stops here".

In my opinion eight years as president is enough and sometimes too much for any man to serve in that capacity. Harry S. Truman

I remember when I first came to Washington. For the first six months you wonder how the hell you ever got here. For the next six months you wonder how the hell the rest of them ever got here. Harry S. Truman

Men make history and not the other way around. In periods where there is no leadership, society stands still. Progress occurs when courageous, skillful leaders seize the opportunity to change things for the better. Harry S. Truman

We shall never be able to remove suspicion and fear as potential causes of war until communication is permitted to flow, free and open, across international boundaries. Harry S. Truman.

This was stated by Harry S. Truman in 1978 and proven to true 33 years later when in 2011 revolutions erupted in the Middle East, fuelled by television, cell phones, internet and technology that could not be imagined in 1978.

GOVERNMENT RESPONSIBILITIES

When the constitution was written, the founders had no inclination as to what life would be like in the 21st century. They thought that travel to the moon was a fantasy that only fiction writers could imagine. We could draw a list of technological achievements that are making the world a better place to live. Few people, particularly our politicians, understand that technology is a double edged sword, on one side making the world a better place to live and on the other side providing us with products that are causing depletion of our mineral resources.

The government of the United States has succeeded in meeting its responsibilities to protect the nation against foreign invaders by building a strong military force to protect not only the United States but also, to protect the security of the world. The constitution did not say that it was our responsibility to protect the world, but we realized that there was no alternative. Prior to World War II our nation wanted to be isolated from the aggressions taking place in Europe, but it was proven that isolation was not a course that we could take.

The government of the United States has succeeded in providing its citizens with the best health care, education, postal service, highways, police, aviation oversight, and social security systems.

Some concerned citizens believe that the United States government has grown too rapidly and is too big. They believe that we should have less regulation and let the market place take care of itself. Very seldom will we hear that the government should be expanded to have more regulation and more oversight. Why? Lobbyists. Is there a single lobbyist who wants more regulation and more oversight? No. If we examine the relationship between corporations, lobbyists and legislators we find that legislators need knowledge and lobbyists are paid to pass on knowledge that is biased in favor of the organization that they represent, usually the large corporations. The legislators, in most cases do not have sufficient knowledge of an issue which puts them in a position of accepting the recommendations of the lobbyists. The legislatures receive campaign contributions from corporations and individuals that are represented by the lobbyists which give them another reason to accept the recommendations of the lobbyists.

Regulation has a bad connotation, thus giving the legislators a reason to vote in favor of deregulation.

Energy was deregulated in 2000 resulting in energy problems throughout the nation, and particularly in California. The ENRON collapse and resulting consequences can be attributed to the lack of regulation and oversight. Legislators made the mistake of deregulation without an increase in oversight. Economists say that when an industry is deregulated there must be a substantial increase in oversight. That was not the case when energy was deregulated.

The airline industry was deregulated in 1978. Prior to that date the Civil Aeronautics Board CAB was the regulating agency controlling routes and fares. The FAA responsibility for safety gave the US the best safety record in the world. Most airlines were making profit and passengers were pleased with service. Why did we deregulate the airline industry? Who wanted the airline industry deregulated? Did the airline industry serve the public better as the result of deregulation? These questions can be answered by asking a person who flew prior to deregulation and is in a position to compare the service before and after deregulation. If that is not enough to judge the effects of deregulation, compare customer satisfaction between US airlines and overseas airlines where governments maintain control of the airlines. Deregulation of the airline industry is another example of a responsibility of the government that they were not able to or willing to fulfill.

The Banking Act of 1933 (Glass–Steagall Act), was a law that established the Federal Deposit Insurance Corporation (FDIC) in the United States and introduced banking reforms, some of which were designed to control speculation. However in 1980, The Depository Institutions Deregulation and Monetary Control Act of repealed provisions of the Glass-Steagall Act that prohibit a bank holding company from owning other financial companies. Then in 1999, the Gramm-Leach-Bliley Act effectively removed the separation that previously existed between Wall Street investment banks and depository banks. Experts believe that this repeal directly contributed to the severity of the financial crisis of 2007–2010 by allowing banks to gamble with their depositors' money. Again, deregulation of the banking industry is an example of a responsibility of the government that they were not able to or willing to fulfill.

The above are just a few of the services and industries that have been deregulated resulting in adverse effects on the economy of the nation.

> *The government of the United States has the responsibility for regulation and oversight of services and industries that are essential to the wellbeing of the nation. When governments deregulate without proper oversight they are shirking their responsibilities.*

Jimmy Carter on energy policies. The 1973 oil embargo showed how vulnerable we were to imported oil from the Middle East. The first president to provide the nation with an energy policy was Jimmy Carter.

Jimmy Carter was president of the United States for one term from January 20, 1977 to January 20, 1981. He suffered at the poles after delivering a televised speech April 18, 1977 on the subject of energy conservation, a subject that Americans did not want to hear. No president prior to or after Jimmy Carter took such a strong position on the looming energy crisis. Many of the predictions from his speech on the energy crisis have proven to be accurate 35 years after his initiative.

By emphasizing conservation rather than the development of new resources, the program seemed to call upon Americans to change their lifestyle, and most did not want to do that. Here is that speech:

> *Tonight I want to have an unpleasant talk with you about a problem unprecedented in our history. With the exception of preventing war, this is the greatest challenge our country will face during our lifetimes. The energy crisis has not yet overwhelmed us, but it will if we do not act quickly. It is a problem we will not solve in the next few years, and it is likely to get progressively worse through the rest of this century. We must not be selfish or timid if we hope to have a decent world for our children and grandchildren. We simply must*

balance our demand for energy with our rapidly shrinking resources. By acting now, we can control our future instead of letting the future control us.

Two days from now, I will present my energy proposals to the Congress. Its members will be my partners and they have already given me a great deal of valuable advice. Many of these proposals will be unpopular. Some will cause you to put up with inconveniences and to make sacrifices. The most important thing about these proposals is that the alternative may be a national catastrophe. Further delay can affect our strength and our power as a nation.

Our decision about energy will test the character of the American people and the ability of the President and the Congress to govern. This difficult effort will be the "moral equivalent of war" -- except that we will be uniting our efforts to build and not destroy. I know that some of you may doubt that we face real energy shortages. The 1973 gasoline lines are gone, and our homes are warm again. But our energy problem is worse tonight than it was in 1973 or a few weeks ago in the dead of winter. It is worse because more waste has occurred, and more time has passed by without our planning for the future. And it will get worse every day until we act.

The oil and natural gas we rely on for 75 percent of our energy are running out. In spite of increased effort, domestic production has been dropping steadily at about six percent a year. Imports have doubled in the last five years. Our nation's independence of economic and political action is becoming increasingly constrained. Unless profound changes are made to lower oil consumption, we now believe that early in the 1980s the world will be demanding more oil that it can produce. The world now uses about 60 million barrels of oil a day and demand increases each year about five percent. This means that just to stay even we need the production of a new Texas every year, an Alaskan North Slope every nine months, or a new Saudi Arabia every three years. Obviously, this cannot continue.

We must look back in history to understand our energy problem. Twice in the last several hundred years there has been a transition in the way people use energy. The first was about 200 years ago, away from wood -- which had provided about 90 percent of all fuel -- to coal, which was more efficient. This change became the basis of the Industrial Revolution. The second change took place in this century, with the growing use of oil and natural gas. They were more convenient and cheaper than coal, and the supply seemed to be almost without limit. They made possible the age of automobile and airplane travel. Nearly everyone who is alive today grew up during this age and we have never known anything different.

Because we are now running out of gas and oil, we must prepare quickly for a third change, to strict conservation and to the use of coal and permanent renewable energy sources, like solar power. The world has not prepared for the future. During the 1950s, people used twice as much oil as during the 1940s. During the 1960s, we used twice as much as during the 1950s. And in each of those decades, more oil was consumed than in all of mankind's previous history. World consumption of oil is still going up. If it were possible to keep it rising during the 1970s and 1980s by 5 percent a year as it has in the past, we could use up all the proven reserves of oil in the entire world by the end of the next decade. I know that many of you have suspected that some supplies of oil and gas are being withheld. You may be right, but suspicions about oil companies cannot change the fact that we are running out of petroleum.

All of us have heard about the large oil fields on Alaska's North Slope. In a few years when the North Slope is producing fully, its total output will be just about equal to two years' increase in our nation's energy demand. Each new inventory of world oil reserves has been more disturbing than the last. World oil production can probably keep going up for another six or eight years. But some time in the 1980s it can't go up much more. Demand will overtake production. We have no choice about that.

But we do have a choice about how we will spend the next few years. Each American uses the energy equivalent of 60 barrels of oil per person each year. Ours is the most wasteful nation on earth. We waste more energy than we import. With about the same standard of living, we use twice as much en-

ergy per person as do other countries like Germany, Japan, and Sweden. One choice is to continue doing what we have been doing before. We can drift along for a few more years. Our consumption of oil would keep going up every year. Our cars would continue to be too large and inefficient. Three-quarters of them would continue to carry only one person -- the driver -- while our public transportation system continues to decline. We can delay insulating our houses, and they will continue to lose about 50 percent of their heat in waste. We can continue using scarce oil and natural [gas] to generate electricity, and continue wasting two-thirds of their fuel value in the process.

If we do not act, then by 1985 we will be using 33 percent more energy than we do today. We can't substantially increase our domestic production, so we would need to import twice as much oil as we do now. Supplies will be uncertain. The cost will keep going up. Six years ago, we paid $3.7 billion for imported oil. Last year we spent $37 billion -- nearly ten times as much -- and this year we may spend over $45 billion. Unless we act, we will spend more than $550 billion for imported oil by 1985 -- more than $2,500 a year for every man, woman, and child in America. Along with that money we will continue losing American jobs and becoming increasingly vulnerable to supply interruptions.

Now we have a choice. But if we wait, we will live in fear of embargoes. We could endanger our freedom as a sovereign nation to act in foreign affairs. Within ten years we would not be able to import enough oil -- from any country, at any acceptable price. If we wait, and do not act, then our factories will not be able to keep our people on the job with reduced supplies of fuel. Too few of our utilities will have switched to coal, our most abundant energy source.

We will not be ready to keep our transportation system running with smaller, more efficient cars and a better network of buses, trains and public transportation. We will feel mounting pressure to plunder the environment. We will have a crash program to build more nuclear plants, strip-mine and burn more coal, and drill more offshore wells than we will need if we begin to conserve now. Inflation will soar, production will go down, people will lose their jobs. Intense competition will build up among nations and among the different regions within our own country. If we fail to act soon, we will face an economic, social and political crisis that will threaten our free institutions.

But we still have another choice. We can begin to prepare right now. We can decide to act while there is time. That is the concept of the energy policy we will present on Wednesday. Our national energy plan is based on ten fundamental principles.

The first principle *is that we can have an effective and comprehensive energy policy only if the government takes responsibility for it and if the people understand the seriousness of the challenge and are willing to make sacrifices.*

The second principle *is that healthy economic growth must continue. Only by saving energy can we maintain our standard of living and keep our people at work. An effective conservation program will create hundreds of thousands of new jobs.*

The third principle *is that we must protect the environment. Our energy problems have the same cause as our environmental problems -- wasteful use of resources. Conservation helps us solve both at once.*

The fourth principle *is that we must reduce our vulnerability to potentially devastating embargoes. We can protect ourselves from uncertain supplies by reducing our demand for oil, making the most of our abundant resources such as coal, and developing a strategic petroleum reserve.*

The fifth principle *is that we must be fair. Our solutions must ask equal sacrifices from every region, every class of people, every interest group. Industry will have to do its part to conserve, just as the consumers will. The energy producers deserve fair treatment, but we will not let the oil companies profiteer.*

The sixth principle, *and the cornerstone of our policy, is to reduce the demand through conservation. Our emphasis on conservation is a clear difference between this plan and others which merely encouraged crash production efforts. Conservation is the quickest, cheapest, most practical source of energy. Conservation is the only way we can buy a barrel of oil for a few dollars. It costs about $13 to waste it.*

The seventh principle is that prices should generally reflect the true replacement costs of energy. We are only cheating ourselves if we make energy artificially cheap and use more than we can really afford.

The eighth principle is that government policies must be predictable and certain. Both consumers and producers need policies they can count on so they can plan ahead. This is one reason I am working with the Congress to create a new Department of Energy, to replace more than 50 different agencies that now have some control over energy.

The ninth principle is that we must conserve the fuels that are scarcest and make the most of those that are more plentiful. We can't continue to use oil and gas for 75 percent of our consumption when they make up seven percent of our domestic reserves. We need to shift to plentiful coal while taking care to protect the environment, and to apply stricter safety standards to nuclear energy.

The tenth principle is that we must start now to develop the new, unconventional sources of energy we will rely on in the next century.

These ten principles have guided the development of the policy I would describe to you and the Congress on Wednesday. Our energy plan will also include a number of specific goals, to measure our progress toward a stable energy system. These are the goals we set for 1985:

-Reduce the annual growth rate in our energy demand to less than two percent.

-Reduce gasoline consumption by ten percent below its current level.

-Cut in half the portion of United States oil which is imported, from a potential level of 16 million barrels to six million barrels a day.

-Establish a strategic petroleum reserve of one billion barrels, more than six months' supply.

-Increase our coal production by about two thirds to more than 1 billion tons a year.

-Insulate 90 percent of American homes and all new buildings.

-Use solar energy in more than two and one-half million houses. We will monitor our progress toward these goals *year by year. Our plan will call for stricter conservation measures if we fall behind. I can't tell you that these measures will be easy, nor will they be popular. But I think most of you realize that a policy which does not ask for changes or sacrifices would not be an effective policy. This plan is essential to protect our jobs, our environment, our standard of living, and our future. Whether this plan truly makes a difference will be decided not here in Washington, but in every town and every factory, in every home and on every highway and every farm. I believe this can be a positive challenge. There is something especially American in the kinds of changes we have to make. We have been proud, through our history of being efficient people.*

We have been proud of our leadership in the world. Now we have a chance again to give the world a positive example. And we have been proud of our vision of the future. We have always wanted to give our children and grandchildren a world richer in possibilities than we've had. They are the ones we must provide for now. They are the ones who will suffer most if we don't act. I've given you some of the principles of the plan. I am sure each of you will find something you don't like about the specifics of our proposal. It will demand that we make sacrifices and changes in our lives. To some degree, the sacrifices will be painful, but so is any meaningful sacrifice. It will lead to some higher costs, and to some greater inconveniences for everyone. But the sacrifices will be gradual, realistic and necessary. Above all, they will be fair. No one will gain an unfair advantage through this plan. No one will be asked to bear an unfair burden. We will monitor the accuracy of data from the oil and natural gas companies, so that we will know their true production, supplies, reserves, and profits.

The citizens who insist on driving large, unnecessarily powerful cars must expect to pay more for that luxury. We can be sure that all the special interest groups in the country will attack the part of this plan that affects them directly. They will say that sacrifice is fine, as long as other people do it, but that their sacrifice is unreasonable, or unfair, or harmful to the country. If they succeed, then the burden on the ordinary citizen, who is not organized into an interest group, would be crushing.

There should be only one test for this program: whether it will help our country. Other generation of Americans have faced and mastered great challenges. I have faith that meeting this challenge will make our own lives even richer. If you will join me so that we can work together with patriotism and courage, we will again prove that our great nation can lead the world into an age of peace, independence and freedom.

The plan was too far ahead of its time for our politicians to comprehend, and certainly too early for the average consumer to understand. There were many private interests opposed to the plan, such as the oil industry which wanted to continue making huge profits by selling oil products without considering the consequences of depleting a nonrenewable natural resource. Much has been said about the looming oil crisis but little action has been taken to prevent depletion of our oil reserves and the economic consequences resulting from a shortage of oil. The large oil corporations and automobile manufacturers had a financial incentive to prevent taking action as proposed in the Jimmy Carter energy proposal so they fought it using a weapon that only large corporations could afford – lobbyists.

LOBBYISTS

Lobbying is considered an essential service providing legislatures with knowledge that is essential for them to properly evaluate new legislation. Lobbying is considered a conduit that provides knowledge of governmental workings to business organizations and also to provide industrial knowledge to governmental employees. Unfortunately, lobbyists are self-serving to the organization that pays their fees with little regard to the general public.

Lobbyists are sometimes referred to as influence peddlers. Lobbying is a legal practice as long as there is no direct payment involved. Influence peddling is the illegal practice of using one's influence in government or connections with persons in authority to obtain favors or preferential treatment for another, usually in return for payment. Also called traffic of influence or trading in influence. In fact, influence peddling is not necessarily illegal as the Organization for Economic Cooperation and Development OECD has often used the term "undue influence peddling" referring to illegal acts of lobbying.

The number of registered lobbyists in Washington has more than doubled since 2000 to more than 34,750. Only a Few other businesses have enjoyed greater prosperity in an otherwise depressed economy. Big-bucks lobbying is luring nearly half of all lawmakers who return to the private sector when they leave Congress. Political historians don't see this as a positive development for democracy. The growth of lobbying makes the balance between those with resources and those without resources worse than it already is.

Corporations need lobbyists to fend off proposals that could restricted them or cost them money. But with pro-business officials running the executive and legislative branches, companies are also hiring well-placed lobbyists to go on the offensive and find ways to profit from the many tax breaks, loosened regulations and other government goodies that increasingly are available.

Leaders of industry are willing to invest money because they see opportunities. They see that they can win things. Washington has become a profit center.

Companies have doubled their lobbying merely to keep track of it all. Much of lobbying today is watching all the change that's going on in Washington. Companies need more people just to stay apprised of what regulators are doing.

Hewlett-Packard doubled its budget for contract lobbyists to $734,000 for the year 2004. Its goal was to pass legislation that would allow the company to bring back to the United States at a dramatically lowered tax rate as much as $14.5 billion in profit from foreign subsidiaries. The extra lobbying paid off. Legislation was approved and Hewlett-Packard will save millions of dollars in taxes. This was an opportunity the business community to make its case with the help of lobbyists and it was successful.

Industry defends its increased use of lobbyists by claiming that as government grows, unless you're right there to limit it, it can intrude in just about any industry. There are agencies that love to do things and acquire new missions. People in industry better have good lobbyists or they're going to get rolled over. But whether it is to protect themselves against harm or to win more benefits, executives and insiders say they have no choice but to hire lobbyists who are deeply rooted in official Washington and its complexities.

The Bush administration has signed into law five major tax-cut bills. His administration has also curtailed regulation. Over the past five years, the number of new federal regulations has declined by 5 percent, to 4,100. The number of pending regulations that would cost businesses or local governments $100 million or more a year has declined even more, by 14.5 percent to 135 over the period.

SUCCESSFUL GOVERNMENT PROGRAMS

The United States is the wealthiest nation in the world. It is also one of the youngest of the developed nations. Some say that the United States had all the elements for success. It had vast unexplored land with unimaginable natural resources waiting to be developed. Because of the political turmoil in other nations, the United States was at the receiving end of ambitious immigrants ready to work hard to achieve success. With all the cards stacked in our favor, success was inevitable.

Our early government saw the potential for success and gave it a little push. Government had a hand in many successful projects such as the national railroad system, Panama Canal, Hoover Dam, the Interstate Highway System, Federal Aviation Administration, plus many social programs and services that most people are unaware of that put the United States far ahead of all nations.

Success of the United States continued until the Great Depression of 1929 which can be blamed on lack of oversight of the stock market by the US government, or lack of oversight of the agriculture industry. We learned from our mistakes, but too slow and not enough.

Recovery from the Great Depression was slow causing many lower class people to suffer until World War II. It is ironic to say that it took World War II with enormous human casualties to bring us out of the Great Depression. It is ironic to say that World War II was a successful government program. Reading history books on how we mobilized for the war proves that it was a successful program.

World War II necessitated production of enormous quantities of war materials. Only essential goods such as food and clothing were produced for the civilian population. Many economists can show statistics proving that during the war, unemployment was practically zero. Not only did we put most of the unemployed to work, but we taught them new skills which they continued to utilize after the war.

New technology learned during the war resulted in new industries which employed many new workers. Although not a new industry, the housing market which came to an abrupt stop because of the war had to provide housing for the GI's returning from the war. Housing projects after the war were numerous and successful until the government decided to interfere by introducing the Community Reinvestment Act which had the best of intentions, but because of lack of government oversight eventually caused the Great Recession of 2008.

It took World War II to get us out of the Great Depression. Now we are in the Great Recession of 2008 and the government does not know what to do. It has tried to correct the situation by throwing money in all directions which has only exasperated the economy.

Our government can learn from our economic success in World War II by declaring war. The United States knows how to declare war on other nations. We have certainly done it enough times. We have declared war on poverty which many considered a joke. We have declared war on drugs and losing that war. Let's declare war on an ominous enemy, an enemy that we can feel and touch, an enemy that is starting to show its ugly head, "ENERGY".   We can win the war on energy if we are willing to heed some of the suggestions in Jimmy Carter's speech of April 18, 1977, or implement some of the ideas in this book.

GOVERNMENT FAILURES

Some of the causes of government failures can be summarize as being caused by complacency, greed, political maneuvering, deregulation, lack of oversight, subsidies, lobbyists, and government give away programs. The largest and most devastating of the failed government programs was the 2008 housing bubble which will continue to have dire consequences on the US economy for decades to come.

The United States government has the responsibility to protect its citizens from financial disasters. Why is the government not capable or willing to perform such duty?

THE GREAT DEPRESSION

The great depression was the result of several negative influences, worst of which was the lack of government oversight of Wall Street. Prices of stocks in the late 1920s went unchecked and could be purchased on margin by putting up less than 2/3 of the value of the stock. The bubble burst on Black Tuesday, October 29, 1929 when stocks plummeted causing the Great Depression.

1. Throughout the 1930s over 9,000 banks failed. Bank deposits were uninsured and thus as banks failed people simply lost their savings. Surviving banks, unsure of the economic situation and concerned for their own survival, stopped making new loans. This exacerbated the situation resulting in little economic growth.

2. With the stock market crash and the fears of further economic woes, individuals from all classes stopped purchasing items. This then led to a reduction in the number of items produced and thus a reduction in the workforce. As people lost their jobs, they were unable to keep up with paying for items they had bought through installment plans and their items were repossessed. More and more inventory began to accumulate. The unemployment rate rose above 25% which meant, of course, even less spending to help alleviate the economic situation.

3. The government learned from the Great Depression and provided regulations designed to prevent reoccurrence of some of the practices that caused the 1929 stock market crash. Those regulations did not anticipate the adverse effects of "products" that wall street geniuses were inventing, such as Hedge Funds, Derivatives, Credit Default Swaps, Mortgage-backed Securities, and Fast-speed Traders that were the underlying cause of the 2008 Great Recession.

THE DOT COM BUBBLE

The dot com bubble started inflating in 1995 when the internet and many information highway startup companies were growing at a rapid rate. Investors got in on the bandwagon buying any stock that had "dot com" after the name. Some of the dot com stocks were selling at price to earnings ratios that could not be justified for 200 years. Wall Street brokers were willing to take money from investors as long as they get their commissions and bonuses. They had no regard for the value of the stocks that they were selling. Mutual fund managers took a similar attitude of buying stocks into their funds as long as they received their commissions. On March 10, 2000 the NASDAQ Composite Index reached a high of 5132 and then started a decline for 3 years reaching a low of 1310. Where was our government and the SEC oversight when the bubble was inflating and Americans were about to lose their life savings?

FAILURE TO PREVENT 9/11

The terrorist attacks on September 11, 2001, could have been prevented if not for intervention by government bureaucracies. There were two documentaries made depicting what went wrong.

"The Man Who Knew" was first presented on PBS Frontline, October 3, 2002. FBI's expert on Al Qaeda, John O'Neill, warned of a threat by Osama ben Laden but was prevented from doing his job investigating terrorists in Yemen by the then United States ambassador, Barbara Bodine who refused to issue him a visa to return to Yemen to complete his work. A repeat of this documentary can be seen in full on the internet under PBS Frontline titled "The Man Who Knew".

Second documentary was produced by the ABC network in a two-night miniseries chronicling the events leading up to the disaster of 9/11, titled "The Path to 9/11". It finally aired after it was heavily edited. This eye-opening documentary exposes the political and media firestorm behind this program.

These documentaries show that United States diplomats tend to forget that they are expected to act in the best interests of their country.

THE HOUSING BUBBLE

The largest failure of government is its involvement in the housing market through the Community Reinvestment Act (CRA) which was enacted by pressure from the Association of Community Organizations for Reform Now (ACORN). The act was politically motivated by legislators who did not understand the economic ramifications. When the housing bubble burst in 2008, it brought down some institutions and threatened other institutions that were considered too big to let fail. The United States government decided that institutions that were too big to fail because they could bring down the entire world economy were bailed out with borrowed money. These bailouts and other fiscal spending caused the national debt to reach the ceiling of $14.294 trillion set by congress on February 12, 2010. After much bickering in Congress on how to increase the national debt, on April 15, 2011, Congress passed the last part of the 2011 United States federal budget. Congressional actions resulted in the United States credit rating to be downgraded from AAA to AA+ resulting in consumer lack of confidence and stock market repercussions.

The subprime mortgage meltdown of 2007 and the pursuing 2008 recession can be traced back to 1970 when Arkansas Community Organization for Reform Now (ACORN) was founded. Its goal was to unite welfare recipients against predatory lending practices in order to obtain affordable housing. While some of Acorn's most notable efforts were in the area of housing, it had numerous health, public safety, education, representation, workers' rights and communications among its victories. Acorn grew beyond Arkansas and changed the name to Association of Community Organization for Reform Now (ACORN). Efforts of ACORN were the driving force behind the CRA of 1977.

The Community Reinvestment Act CRA of 1977 was intended to reduce discriminatory credit practices against low and moderate neighborhoods so that certain neighborhoods could have improved development. The Community Reinvestment Act of 1977 seeks to address discrimination in loans made to individuals and businesses from low and moderate-income neighborhoods. The Act mandates that all banking institutions that receive Federal Deposit Insurance Corporation (FDIC) insurance be evaluated by Federal banking agencies to determine if the bank offers credit in all communities in which they are chartered to do business. Federal regulations dictate agency conduct in evaluating a bank's compliance in five performance areas, comprising twelve assessment factors. This examination culminates in a rating and a written report that becomes part of the supervisory record for that bank. Compliance does not require institutions to make high-risk loans that may bring losses to the institution. Compliance with this requirement is has been very controversial. An institution's CRA compliance record is taken into account by the banking regulatory agencies when the institution seeks to expand through merger, acquisition or branching. CRA resulted in sub-prime mortgages being given to persons that presented a high risk of default. The resulting easy mortgages stimulated the housing market by requiring government sponsored enterprises, Fannie Mae and Freddie Mac, to buy those low quality mortgages from the banks, guarantee them and then sell them to the public.

Sub-prime mortgages were originated by large banks such as Countrywide Financial, Washington Mutual, Bank of America and many other smaller institutions. In 2004, the Federal Bureau of Investigation warned of an epidemic in mortgage fraud, an important credit risk of nonprime mortgage lending, which, they said, could lead to a problem that could have as much impact as the S&L crisis of the 1980 The housing bubble created by risky sub-prime mortgages was doomed to burst, and in 2007, it started to collapse.

As the housing market bubble started to burst, the public learned that financial institutions were using risky financial derivatives such as Credit Default Swaps and Collateralized Debt Obligations. In spite of the fact that the government was warned about risky derivatives, they refused to take action to regulate the practices. October 20, 2009, PBS Frontline presented the documentary "The Warning" revealing the involvement of Brooksley Born, Larry Summers, Alan Greenspan and Robert Rubin in dangerous financial derivatives. "We didn't truly know the dangers of the market, because it was a dark market," says Brooksley Born, the head of an obscure federal regulatory agency -- the Commodity Futures Trading Commission [CFTC] - who not only warned of the potential for economic meltdown in the late 1990s, but also tried to convince the country's key economic powerbrokers to take actions that could have helped avert the crisis. "They were totally opposed to it," Born says. "The *Warning"* unearths the hidden history of the nation's worst financial crisis since the Great Depression. Brooksley Born, who speaks for the first time on television about her failed campaign to regulate the secretive, multitrillion-dollar derivatives market which helped trigger the financial collapse in the fall of 2008. "I didn't know Brooksley Born," says former SEC Chairman Arthur Levitt, a member of President Clinton's powerful Working Group on Financial Markets. Greenspan, Rubin and Summers ultimately prevailed on Congress to stop Born and limit future regulation of derivatives. "Born faced a formidable struggle pushing for regulation at a time when the stock market was booming," Kirk says. "Alan Greenspan was the maestro, and both parties in Washington were united in a belief that the markets would take care of themselves."

Now, many of the same men who shot down Born now hold key positions in the Obama administration, The Warning reveals the complicated politics that led to this crisis and what it may say about current attempts to prevent the next one. "It'll happen again if we don't take the appropriate steps," Born warns. "There will be significant financial downturns and disasters attributed to this regulatory gap over and over until we learn from experience."

Repercussions from the housing bubble were being felt more than three years later when on August 8, 2011, AIG announced that it was suing Bank of America for more than $10 billion, alleging that BofA, and its acquisitions Merrill Lynch and Countrywide Financial, participated in "massive fraud" when they sold mortgage-backed securities to AIG between 2005 and 2007. AIG says that more than 40 percent of the mortgages were presented as being more secure than they actually were.

As the housing market collapsed in 2008, the United States government took the position that some institutions were too big to fail and took action to bail them out. The Emergency Economic Stabilization Act of 2008 enacted October 3, 2008, commonly referred to as a bailout of the U.S. financial system, is a law enacted in response to the subprime mortgage crisis authorizing the United States Secretary of the Treasury to purchase distressed assets, especially mortgage-backed securities, and make capital injections into various financial institutions. Both foreign and domestic banks were included in the program. The bailout programs added to the already large national debt.

December, 2009, the debt ceiling of $12.1 trillion was exceeded. Congress has raised the debt ceiling four times in the past two years and will probably have to do it again. Lawmakers really don't have a choice but to raise the ceiling and they know it. Put simply, if they don't raise the ceiling, the country will go into default on its debt. The domino effect would be painful, to say the least.

Increase in the debt limit was negotiated in Congress in July 2011. Failure to raise the limit by August at the latest would disrupt financial markets and further endanger the recovery from the housing crisis. Political negotiations centered around deep spending cuts and unwise tax concessions, claiming that either one could damage the economy. A compromise bill was signed at the 11th hour. Because of the inability of the United States Congress to pass a debt limit bill in a timely manner, confidence in financial obligations of the United States was compromised.

This is an example of the US Government getting involved in a program that went terribly wrong. Neither the government nor the SEC were able to foresee the bubble bursting or properly control the damage.

PONZI SCHEMES

Ponzi schemes are named after Charles Ponzi, who in the 1920s lured investors with promises of profits that could not be obtained in a legal business. All Ponzi schemes have one thing in common. Eventually, the bubble must burst. Largest and most famous of the Ponzi schemes is Bernard L. Madoff Investment Securities which collapsed on December 10, 2008. Many articles are available describing the details of the scheme and the lack of government intervention. The SEC was warned of the Madoff scheme, but took no action to protect investors.

This is another example of the US Government failing to provide oversight and take action to protect its citizens from a bubble that eventually burst.

SOLYNDRA BANKRUPTCY

March 19, 2009, the White House announced a loan to Solyndra of Fremont, California for $535,000,000 to build a so-called "green energy" industry that creates jobs and safeguards the environment. The Solyndra loan was so central to this strategy that the administration had Obama personally announce it, and later sent the president to the company's solar panel manufacturing facility in Fremont, California to celebrate its work.

Solyndra was a manufacturer of panels of CIGS thin-film solar cells based in Fremont, California. Once touted for its unusual technology, plummeting silicon prices led to the company being unable to compete with more conventional solar panels. On September 1, 2011, the company ceased all business activity, filed for Chapter 11 bankruptcy, and laid off all employees.

The Energy Department knew Solyndra LLC was violating terms of its U.S. loan by December and agreed to keep providing taxpayer support for the solar-panel maker that filed for court protection nine months later.

Solyndra President and CEO Brian Harrison, along with William Stover, the company's CFO, refused to answer questions at a House Energy and Commerce Committee hearing, citing their Fifth Amendment protection against self-incrimination.

Here are some facts and figures regarding solar energy based on reports by the United States Energy Information Administration, table 1.3, as of 2007. Combined solar heat and solar photovoltaic energy production in the United States was .081 quadrillion BTU per year. Total energy consumption in the United States was 101.445 quadrillion BTU per year. These figures show that as of 2007 all forms of solar energy account for less than 0.1% of the total energy consumed in the United States. We were barking up the wrong tree.

We haven't yet declared war on energy, but we're fighting the wrong battle and losing that battle. Add this to the list of government failures.

FINANCIAL BUBBLES

Financial bubbles are the result of banks and financial institutions borrowing, referred to as leverage, to finance their investments. For example, borrowing to finance investment in the stock market became increasingly common prior to the Wall Street Crash of 1929. If a financial institution borrows in order to invest more, it can potentially earn more from its investment, but it can also lose more than all it has. Therefore leverage magnifies the potential returns from investment, but also creates a risk of bankruptcy. Since bankruptcy means that a firm fails to honor all its promised payments to other firms, it may spread financial troubles from one firm to another. The average degree of leverage in the economy often rises prior to a financial crisis. Leverage of money by banks is not only a legal means of doing business, but it is made possible and encouraged by the US Federal Reserve System.

GAMBLING IN THE FINANCIAL INDUSTRY

A person works forty to fifty years so that when he retires there will be money that he can use to live comfortably. If he puts that money in a savings account in a local bank, it will lose its value due to taxes and inflation that is built into our economy by the United States Federal Reserve System as a means of wiping out federal debt. The United States government does not provide its citizens with a safe method of protecting its money against inflation.

The alternative to losing money in a bank saving account is to invest it in the stock market. When putting money into the stock market, there is a fine line between investing and gambling. The United States government has allowed Wall Street to become a means of legitimate gambling by allowing "products" such as Hedge Funds, Derivatives, Credit Default Swaps, Mortgage-backed Securities, and Fast-speed Traders. As the result of these tools that are permitted on Wall Street, many billionaires have been created at the expense of many hard working citizens who lost money that would have allowed them to achieve their dream of comfortable retirement.

To get the detailed story about how "derivatives" contributed to the 2008 financial melt-down and how it could have been averted, go to the internet and watch a rerun of "The Warning" which was first presented on PBS Frontline, October 20, 2009.

CREATING JOBS

Politicians give us rhetoric about how they have or how they will create jobs, but they do little more than throw money at the economy. Politicians debate the direction that the money will be thrown and how much will be thrown.

Our government has the power to finance large projects that will build our economy, alleviate our energy crisis, and at the same time put the unemployed to work. A single word can describe these projects is, "Electrification". Technically, we know how to do it, but financially, we are trapped by government politics.

We must electrify our railroads, our highways, our automobiles and our trucks. Statistics show that we making progress in this direction, but too slow when compared to the rate that we are depleting our oil reserves.

During the Great depression, the United States proved that we could create jobs by defining projects that were needed by our nation. These were large projects that employed millions and made life better for millions.

In 1935 the Rural Electric Administration (REA) was created to bring electricity to rural areas like the Tennessee Valley. The Roosevelt Administration believed that if private enterprise could not supply electric power to the people, then it was the duty of the government to do so.

During the Great Depression, we built the Hoover Dam, started the Tennessee Valley Authority, and the Works Progress Administration. They helped, but the project that got us out of the Great Depression was World War II.

If it takes a war to mobilize us against an enemy, let's declare war on energy and hope that we learned our lessons from our war on drugs and our war on poverty.

MEDICARE REFORM

Socialized medicine is a term used to describe a system for providing medical and hospital care for all at a nominal cost by means of government regulation of health services and subsidies derived from taxation. The term Socialized Medicine is avoided in political debates concerning health care, because of the U.S. culture's historically negative associations with socialism. The term was first widely used in the United States by advocates of the American Medical Association in opposition to President Harry S. Truman's 1947 health-care initiative.

Socialized medicine is available in many developed countries, including United Kingdom, Canada, Australia, China, Cuba, Finland, Israel and Russia, where medical care financed by the government is available to all citizens. Quality of socialize medicine in other countries has been criticized by Americans, but a better system has not been made available to Americans. Why?

Waste by doctors, waste by patients, abuse by the legal profession, greed by drug companies, and fraud by some doctors. The list goes on and on.

In theory, the United States has a good system that needs repair: not a new system. Our political leaders are lawmakers, not repairmen with the ability to fix our system without first destroying the system and our economy.

NATIONAL DEBT CRISIS

Political turmoil within political parties caused the credit rating of the United States to be lowered one level from AAA to AA+ while keeping the outlook at "negative" as it becomes less confident Congress will end tax cuts or change entitlements. The rating may be cut to AA within two years if spending reductions are lower than agreed.

Lawmakers agreed on August 2, 2011 to raise the nation's $14.3 trillion debt ceiling and put in place a plan to enforce $2.4 trillion in spending reductions over the next 10 years, which is less than the $4 trillion S&P preferred. Agreement by lawmakers came too little and too late. It was not the final agreement as much as the difficulty out lawmakers had in reaching the agreement that caused the credit rating to be lowered.

TELEVISION DOCUMENTARIES

Twenty first century communications reveal many of the short-comings of mankind such as inequalities between nations, inequalities of individuals within a nation, differences in religious beliefs. Modern communications by television and the internet allow viewers to sit back in an easy chair and see all the shortcomings of our leadership, and do nothing about them.

Several documentaries on government failures have been presented by PBS Frontline and other private media. Complete documentaries can be viewed on the internet. Here are a few.

THE MAN WHO KNEW was first presented by PBS Frontline on October 3, 2002. John O'Neill, the FBI's expert on Al Queda warned of its threat, but his maverick style doomed his career.

O'Neill would probably be no more memorable than any other FBI agent, except for a man named Osama Bin Laden. O'Neill had partaken in the investigation of almost every America related terrorist event in recent history, including Oklahoma City, the USS Cole in Yemen, the embassy bombings in Africa, and the bombing of the Khobar Towers in Saudi Arabia. Yet what made O'Neill noteworthy was the fact he fingered Osama Bin Laden and Al Qaeda as a terrorist threat to the United States in 1995, six years before the infamous attack on the World Trade Center. However, just like a painter without a canvas, O'Neill was a man with an answer to a question no one had the foresight to ask, and no one would listen until it was too late.

A revealing story of John O'Neill is presented in the book, The Man Who Warned America: The Life and Death of John O'Neill, the FBI's Embattled Counterterror Warrior by Murray Weiss (Jun 1, 2004).

THE WARNING, first presented on PBS Frontline, October 20, 2009 unearths the hidden history of the nation's worst financial crisis since the Great Depression. At the center of it all is Brooksley Born, who speaks for the first time on television about her failed campaign to regulate the secretive, multitrillion-dollar derivatives market whose crash helped trigger the financial collapse in the fall of 2008.

"I didn't know Brooksley Born," says former SEC Chairman Arthur Levitt, a member of President Clinton's powerful Working Group on Financial Markets. "I was told that she was irascible, difficult, stubborn, unreasonable." Levitt explains how the other principals of the Working Group -- former Fed Chairman Alan Greenspan and former Treasury Secretary Robert Rubin -- convinced him that Born's attempt to regulate the risky derivatives market could lead to financial turmoil, a conclusion he now believes was "clearly a mistake." Greenspan, Rubin and Summers ultimately prevailed on Congress to stop Born and limit future regulation of derivatives. "Born faced a formidable struggle pushing for regulation at a time when the stock market was booming, Alan Greenspan was the maestro, and both parties in Washington were united in a belief that the markets would take care of themselves."

Now, with many of the same men who shut down Born in key positions in the Obama administration, The Warning reveals the complicated politics that led to this crisis and what it may say about current attempts to prevent the next one. "It'll happen again if we don't take the appropriate steps," Born warns. "There will be significant financial downturns and disasters attributed to this regulatory gap over and over until we learn from experience."

INSIDE JOB is a documentary that can be viewed online. It deals with the subject of how and why the housing market had a meltdown which caused a financial crisis. This documentary paints a galling portrait of an unfettered financial system run amok without accountability. It describes the systemic corruption of the United States by the financial services industry and the consequences of that systemic corruption. The film explores how changes in policy and banking practices helped create the financial crisis.

INSIDE THE MELTDOWN presented by PBS Frontline, February 17, 2009 could have been titled, "Too Big to Fail". As the housing bubble burst and trillions of dollars' worth of toxic mortgages began to go bad in 2007, fear spread through the massive firms that form the heart of Wall Street.

By the spring of 2008, burdened by billions of dollars of bad mortgages, the investment bank Bear Stearns was the subject of rumors that it would soon fail. Bear Stearns issued huge amounts of asset-backed securities. As investor losses mounted in those markets in 2006 and 2007, the company actually increased its exposure, especially the mortgage-backed assets that were central to the subprime mortgage crisis. In March 2008, the Federal Reserve Bank of New York provided an emergency loan to try to avert a sudden collapse of the company. The company could not be saved.

The collapse of the company was a prelude to the risk management meltdown of the Wall Street investment bank industry in September 2008, and the subsequent global financial crisis and recession.

Within days, in September 2008, another investment bank, Lehman Brothers, was on the brink of collapse. Once again, there were calls to bail out the Wall Street giant but because of political pressure the company was left to fail.

THE MADOFF AFFAIR on PBS Frontline, May 12, 2009 unearthed Madoff's operation, the film ultimately takes viewers inside the heart of the fraud. Rich Caputo, who maintained the computers and printers at Madoff Securities, saw the volume of paperwork that went into sustaining the fiction. "We were spitting out statements constantly," Caputo tells Frontline. "The fact that all of these statements were just sort of made out of thin air is pretty shocking."

Madoff was repeatedly investigated by the SEC, but it found nothing amiss. Madoff remained a pillar of the financial world until the markets came crashing down in late 2008. "The fact is, he easily could have gone through his life without this being found out," says Madoff investor Burt Ross. "The only reason that this ended was because, at one given point in time, the economy did so badly that people wanted and needed to get money out of Madoff's investments." It was not the SEC that uncovered the Ponzi Scheme.

THE SPILL on PBS Frontline, October 26, 2010 reveals the events leading up to the Deepwater Horizon drilling rig explosion on April 20, 2010. The explosion and subsequent fire on the Deepwater Horizon semi-submersible Mobile Offshore Drilling Unit (MODU), which was owned and operated by Transocean and drilling for BP in the Macondo Prospect oil field about 40 miles

(60 km) southeast of the Louisiana coast. The explosion killed 11 workers and injured 16 others; another 99 people survived without serious physical injury. It caused the Deepwater Horizon to burn and sink, and started a massive offshore oil spill in the Gulf of Mexico. This environmental disaster is now considered the second largest in U.S. history, behind the Dust Bowl. It shows how, over the past decade, BP vaulted from an energy "also-ran" to one of the biggest companies in the world, gobbling up competitors in a series of mergers that delivered handsome profits for shareholders. The investigation by Frontline and the nonprofit newsroom ProPublica shows that BP's leadership failed to create a culture of safety in the massive new company. As BP took increasingly big risks to find oil and extract it, the company left behind a trail of mounting problems: deadly accidents, disastrous spills, countless safety violations. Each time, BP acknowledged the wider flaws in its culture and promised to do better. The rhetoric was empty. From the refineries to the oil fields to the Gulf of Mexico, BP workers understood that profits came first.

U.S. Minerals Management Service had the responsibility to oversee the drilling rigs in the Gulf of Mexico but did not perform their duties. Were they pressured to look the other way so that the oil companies could "drill baby, drill, drill, drill"?

THE STORM first reported on PBS Frontline, November 22, 2005 examines the chain of decisions that slowed federal response to the calamity in New Orleans, and the government's failure to protect thousands of Americans from a natural disaster that long had been predicted, and the state of America's disaster-response system.

Valuable time was lost negotiating over who should take charge and talks about some of the frustrating meetings. There was a dance going on about who had ultimate authority. "The Storm" also examines the state of America's disaster-response system and whether we have the ability to handle future catastrophic events in light of the failures with Hurricane Katrina. The U.S. still doesn't have a resilient and interoperable communications system in place for first responders in cities and metropolitan areas. Officials point to state and local government opposition to having federally man-dated standards for communication systems and other homeland security equipment. As a result, we have cities that bought air-conditioned garbage trucks with homeland security money without ever solving their communications problems."

BLIND SPOT is a documentary that illustrates the current oil and energy crisis that our world is facing. Whatever measures of ignorance, greed, wishful thinking, we have put ourselves at a crossroads, which offer two paths with dire consequences. If we continue to burn fossil fuels we will choke the life out of the planet and if we don't our way of life will collapse. This documentary makes "An Inconvenient Truth" look like a sitcom.

HEAT on PBS Frontline, October 21, 2008. Our wealth, our society, our being is driven by oil and carbon. It's intellectually dishonest to somehow say we can get some new type of light bulbs or drive a Prius, and our energy problems will disappear. No, this is going to take massive technological innovation. It's going to take changes in the way we live and work. And it's going to take cooperation of unprecedented degrees among business and government and among countries. That's where we are, and that's why there's no other word except `daunting. One has to approach this with great humility. We seem incapable of grasping what's at stake here, perhaps because so much is at stake. Addressing this really means reinventing the engine of our lives—which is fossil fuels."

POISONED WATERS on PBS Frontline, April 21, 2009. More than three decades after the Clean Water Act, iconic American waterways like the Chesapeake Bay and Puget Sound are in perilous condition and facing new sources of contamination.

With polluted runoff still flowing in from industry, agriculture and massive suburban development, scientists note that many new pollutants and toxins from modern everyday life are already being found in the drinking water of millions of people across the country and pose a threat to fish, wildlife and, potentially, human health.

FUTURE FAILURES OF THE UNITED STATES GOVERNMENT

Oil depletion is showing its ugly head in many ways. Higher gasoline prices are upon us, but most consumers do not want to face the real reason for the increasing prices, which in very simple terms is the fact that oil reserves are becoming depleted resulting in new oil production from wells that are more expensive to drill. Consumers blame the high prices for gasoline on Wall Street speculators and enormous profits being made by the big oil companies. High oil prices and high gasoline prices are caused by the simple and well-known economic expression of "Supply and Demand". The world supply of oil is becoming depleted at a faster rate than the world is able to or willing to reduce its demand for oil.

There have been many television documentaries, magazine articles and books written on the subject of oil depletion. Several "Doom and Gloom" are listed in the appendix of this book. Add this book to the list.

Our political leaders and consumers do not want to hear the undisputed fact that oil and all the consumer products dependent on oil, including basic necessities such as food will become increasingly more expensive to the point that our economy will be drastically effected.

Consumers do not want to hear that their life styles must change drastically in order to prevent an economic disaster, and political leaders certainly do not want to tell them that their life styles must change.

Our political leaders have been told that our oil supplies are becoming depleted, but they do not fully understand or do not want to understand the gravity of the situation. The only political leader who understood the looming energy situation and was willing to take action was Jimmy Carter. At the beginning of this section is a transcript of the speech he made to the nation on April 18, 1977. The public did not want to hear what he had to say. Historians claim that because of that speech, Jimmy Carter was not reelected to a second term as President.

Oil depletion and the national debt of the United States are treated by politicians and economists as unrelated issues. The section titled "Economics and Energy" shows that all aspects of our economy are dependent upon energy, and as the cost of energy increases, our economy will slow down.

WHAT CAN BE DONE?

We can leave the problem in the hands of our political leaders to resolve which means taking little or no action. With little action by our political leaders, we will wait until the price of oil reaches a level that brings down our economy and we enter a depression. After entering a depression caused by depletion of oil reserves, there will be less demand for the remaining high priced oil and the situation will correct itself. That is the way a national or world economy works. Can our politicians comprehend what the humanitarian consequences of such a correction will be? Answer is NO.

Industry leaders see that fossil fuels are becoming depleted and that oil is leading the race to depletion. Industry is merely delaying the eventual depletion of oil by slowly switching our energy needs to more plentiful coal and natural gas.

Industry leaders are aware that there will be a greater impact on our environment as we switch from oil to more plentiful coal and natural gas. Coal is polluting our atmosphere with CO_2 and natural gas extraction is polluting our drinking water supplies with numerous toxins when hydraulic fracturing is used to extract natural gas.

When availability of an essential commodity is insufficient to meet the needs of all persons there will be rationing in one form or another. The first form of rationing is to allow those with sufficient money to purchase the commodity in whatever quantity they can afford and squander the commodity in whatever manner they chose. The second form of rationing is to allocate the commodity in an equitable manner similar to the method used during World War II, but using technology available in the 21st century. Enforcement of equitable rationing of a commodity has always been considered political suicide which leaves rationing to be determined by the ability to pay. The consequences of not having an equitable form of distribution of an essential commodity can only lead to consumer unrest with the possibility of revolt, starvation, suffering and eventually collapse of an economic system.

Rationing will not come from our political or industrial leaders. It will come from the consumers who see it as the best solution and at the same time an equitable solution. A practical rationing system using modern technology is discussed in Section Four, Economics of Energy.

SECTION FOUR
ECONOMICS OF ENERGY

Economics is one of the most difficult subjects to understand. It is far from being an exact science with many theories, most of which have never been proven.

After the 2008 housing market bubble burst and a financial crisis ensued, practically all of our economic leaders admitted that they did not fully understand some of the financial instruments that caused the crisis. These included some of the top economic brains of the United States including former SEC Chairman Arthur Levitt, former Fed Chairman Alan Greenspan, former Treasury Secretary Robert Rubin, and former Treasury Secretary Larry Summers. They admitted that they did not understand the derivatives market which ultimately was a key factor in the 2008 financial meltdown. Alan Greenspan was the maestro who believed that the markets would take care of themselves. He later admitted that was wrong in his understanding of the derivatives market.

Many of the same men who later admitted that they did not understand the derivatives market were appointed to key positions in the Obama administration with nothing being done about the derivatives market. It can happen again if we don't take appropriate steps. There will be significant financial downturns and disasters attributed to this regulatory gap over and over again until we learn how to deal with our mistakes.

Why are the men who admitted that they did not understand the derivatives market still in key positions in the Obama administration? Answer is that the financial markets are very complex and these are the only men who are capable of dealing with such a complex market, which poses the next question. Why aren't the financial markets made less complex by preventing derivatives from rearing their ugly head again? Answer is political pressure from lobbyists working on behalf of giant Wall Street firms.

Economics covers the broadest spectrum of disciplines including banking, insurance, stocks, bonds, interest rates, and energy. Why energy? If there is not enough energy to drive the industries of a nation to produce goods and services, the economy of that nation will be adversely affected. The 21st century started to see the adverse effects of increased energy costs without knowing how to deal with the problem.

There have been many economists holding high level positions in the United States government guiding the fortunes of the nation and its citizens who did not fully understand the signals of downturn. One of the most famous economics experts of the 21st century was Allan Greenspan, chairman of the Federal Reserve for 18 years before he stepped down in January 2006. Almost three years after stepping down as Chairman of the Federal Reserve, on October 23, 2008, at a congressional hearing, he admitted he made a mistake by putting too much faith in the self-correcting power of free markets and that he had failed to anticipate the self-destructive power of wanton mortgage lending. He admitted that he didn't understand what was happening. There were other economic leaders who did not fully understand the dangers of derivatives and credit default swaps, some of whom remain as top financial advisors to President Obama.

United States top financial leaders are not the only ones who did not understand derivatives and credit default swaps. September 16, 2008 the world's largest insurer and the 18th largest company in the world, American International Group was brought to the brink of bankruptcy because of the credit default swaps that it issued. The AIG Financial Products division headed by Joseph Cassanogiant wall , had entered into those credit default swaps to insure $441 billion worth of securities in the sub-prime housing market. When the housing market collapsed it took down AIG. Cassano was named as one of the "Ten Most Wanted: Culprits" causing the 2008 financial collapse in the United States.

Where was the CEO of American International Group when those credit default swaps were issued? Did he know what the risks were when he approved the issuance of those credit default swaps? Answer – NO. Is the CEO of a company expected to understand all of the financial instruments that the company deals with? Answer - YES. Is it humanly possible for any one person to understand all of the financial instruments that a company the size of AIG deals with? Answer is NO.

Because of the complexity of making a decision in this world of multi-national corporations and modern technology, corporate leaders will make decisions by taking a course of action based on a calculated risk which is a form of gambling or taking the easy path by accepting the recommendations of subordinates who may not be properly qualified which results in making imprudent decisions. Whatever it is called, leaders of government and industry do it.

Conclusion – some of the economic issues facing our leaders in this modern world of technology are too difficult for even the CEO of a large corporation or the best brains of the United States government to understand.

> ***The only way that we can make national and international economics understandable by the average person and politician is to eliminate the complex financial instruments such as derivatives and credit default swaps. Would we have a better world without them? YES***

RETAIL PRICE OF ENERGY

Ask any driver what they paid for gasoline and they will say that the last time they filled up their car they paid $3.29 per gallon, and probably make a comment about what they paid a year ago, and that the government should do something about the high price of gasoline. Ask that person what they paid per BTU and they will comment that they have no idea. How many BTU's did you use going to Wal-Mart yesterday? Are you kidding?

Ask that same person what they paid for electricity and they will probably say that their electric bill last month was $237. What did you pay per kilowatt hour? What did you pay per BTU? Answer will probably be the same, that they have no idea.

Ask that person what part of their total income goes to pay for energy, and the same answer. What is the largest corporation? Wal-Mart. Correct. What is the second largest corporation, the third, the fourth, the fifth? Probably the oil companies. Correct.

There are a lot of questions that you could ask the average consumer about energy and you will get a lot of I don't know answers. This book will make the average consumer better aware of the present energy situation and the looming energy crisis that we face in the future.

COST OF PRODUCING OIL

Depletion of fossil reserves is inevitable. Oil depletion will occur in the 21st century with many experts claiming that oil depletion has already started as measured by decreasing world production. For this reason the cost of producing oil should be studied before studying natural gas and coal.

During the 2012 republican nomination campaign for president, one of the candidates promised that if elected he would bring the price of gasoline back to $2.50 per gallon. He never said how he would fulfill such a promise. Would it be temporary by releasing all the oil in the nation's strategic oil reserves? Would it be by subsidizing the cost of gasoline? Would it be by rationing gasoline? Would it be by going to war and stealing oil from other nations? Did that candidate review the speech presented by President Jimmy Carter on April 18, 1977? Probably not. That candidate did not understand the logic of why the price of oil has been increasing over time.

Here is simple logic that explains why the cost of oil has been increasing over time.

1.   CONSUMPTION of oil is increasing because technology is giving us more uses for oil. Increases in consumption of oil per capita multiplied by increases in population account for the large increase in demand for oil. Some new uses of oil improve our health, and wellbeing. However, some new uses of oil cause extravagant consumption by toys such as gas guzzling cars, boats and airplanes. Vacation and recreation travel consume large amounts of oil for commercial airplanes and cruise ships. Consumption will continue to increase until the cost of oil reaches a point where the middle class cannot afford basic necessities such as food. Will sufficient renewable energy be developed in time to prevent widespread suffering and starvation? Probably not. Will rationing be put in place in time to prevent widespread suffering and starvation? Probably not.

2. RESERVES of oil are being depleted at a rate greater than new oil reserves are being discovered. Fossil reserves are not being replenished by nature. Remaining reserves are being discovered in areas that make them more difficult and expensive to produce. Again, will renewable energy be developed in time to prevent widespread suffering and starvation? Probably not.

3. PRODUCTION of oil is becoming more difficult and more expensive. The following paragraphs show why.

Oil was first discovered in the 13th century as a messy substance that seeped to the surface. It was simply scooped up and used for lighting and medicine. In the 16th century it was used for street lighting. As more uses were found for oil, man started to dig simple shallow wells to obtain oil. Oil was plentiful and cheap.

In 1850 oil was distilled to obtain kerosene which was used in lamps to obtain provide clean burning light. The industrial revolution placed more dependence on oil to drive the machines, and in 1895 the invention of the internal combustion engine required more oil to obtain gasoline for the engines. Oil was no longer plentiful and cheap. Man had to look for it by digging and drilling in areas where oil had seeped to the surface.

In 1910 drilling for oil in California produced the Lakeview gusher near Los Angles, California. It had a rate of 100,000 barrels of oil per day. It was a hit and miss technology with many misses. Enough oil was located to meet the needs of the industrial revolution. It was again plentiful and cheap.

In 1920 the industrial revolution and mass production of automobiles placed more demand on oil. It became necessary to locate more oil to meet the new demands. In order to locate enough oil, technology using seismic waves made it possible to locate more oil. Oil remained plentiful and cheap.

In 1934 using seismic waves, oil was located in areas below the sea and was obtained using floating drilling rigs. Because of these new technologies, oil remained plentiful and cheap.

In 1967 oil sands in Alberta, Canada started to produce oil. There was plenty of it but it was more expensive to produce. We could still say that oil was plentiful, but we could no longer say that it was cheap.

In 1968 oil was discovered on the North Slope of Alaska, and the North Sea. Producing oil from these remote areas proved that oil was no longer plentiful and certainly not cheap.

The 1973 oil embargo showed how vulnerable the United States was to imported oil. Little was done by our leaders to change the situation.

On April 18, 1977, President Jimmy Carter gave a televised speech on the subject of energy conservation. Many of his predictions on the looming energy crisis proved to be accurate 35 years after giving his speech.

From 1979 to 1981 oil prices rose from $13.00 to $34.00 a barrel. The increase in prices was a combination of increased production costs and the political situation in the Middle East. The political situation in the Middle East will change with time but the cost of producing oil will continue to become more expensive.

After July 6, 1988 the cost of producing oil had to be figured not only in terms of dollars, but in terms of human lives lost to produce oil. That was the day that a large scale explosion occurred on the Piper Alpha North Sea oil and gas production platform. The resulting fire destroyed the platform, killing 167 men, with only 59 survivors.

On April 20, 2010 the Deepwater Horizon rig, in the Gulf of Mexico, killed 11 workers and causing a major environmental catastrophe along the Gulf Coast

On April 19, 2011 a blow out in a shale gas well owned by Chesapeake Energy in Pennsylvania ignited debate on the safety and environmental impact of hydraulic fracturing

These recent incidents show that the cost of oil must be measured not only in dollars, but in terms of human lives and environmental damage.

VARIATIONS IN THE COST OF PRODUCING OIL

	Estimated Production Costs per Barrel in 2008 dollars
Middle East – N. African oilfields	6 - 28
Other conventional oilfields	6 - 39
CO2 enhanced oil recovery	30 80
Deep water oilfields	32 - 65
Enhanced oil recovery	32 - 82
Arctic oilfields	32 – 100
Heavy oil/bitumen	32 – 68
Oil shale	52 – 113
Gas to liquids	38 – 113
Coal to liquids	60 - 113

INTERNATIONAL ENERGY AGENCY – 2008 WORLD ENERGY OUTLOOK

From the above it can be seen that the cost of oil will increase as it becomes more expensive to extract it from the earth. However, in the case of Middle East oil, the cost of extracting it from the earth will remain low because the oil flows freely. Geologists believe this is partially due to heat from the desert maintaining the oil at low viscosity and easy to produce. The other explanation is the pressure that forces it to the surface. This can be understood by observing the oil fires after Saddam Hussein was defeated in the Gulf War and he ordered the Kuwait oil wells set on fire. The fires burned out of control for weeks because the oil was being forced to the surface by internal pressure.

Photos of oil wells in the continental United States show wells with pumps (pumpjacks) that are necessary to bring the oil to the surface. Historically, in the United States, some oil fields existed where the oil rose naturally to the surface, but most of those fields have long since been used up, except in certain places in Alaska. Production from these wells is more expensive than the free flowing oil from wells in the Middle East.

Oil sands have recently been considered to be part of the world's oil reserves. Higher oil prices and new technology enable them to be profitably extracted. They are often referred to as unconventional oil in order to distinguish from the free-flowing hydrocarbon mixtures known as crude oil produced from oil wells.

Although profitable, oil produced from oil sands is the most expensive oil and is a factor in setting 21st century oil prices.

VARIATIONS IN THE PRICE OF OIL

Why are there large variations in the price of crude oil and the resulting price of gasoline at the? The answer is complex involving the cost of producing oil, financial speculators, goal conflicts, and price manipulation by the oil producers. There is no one simple answer.

The graph below (solid line) shows average annual prices per barrel of crude oil. During any year there may be larger variations that do not show in the chart due to annual averaging.

The dotted line below is an attempt to show what the price of oil would be if not for variations in the cost of producing oil, fluctuations due to financial speculators, global conflicts and price manipulation by oil producers.

If the United States or any other nation could eliminate its dependence on foreign oil, these factors would be eliminated and prices would stabilize.

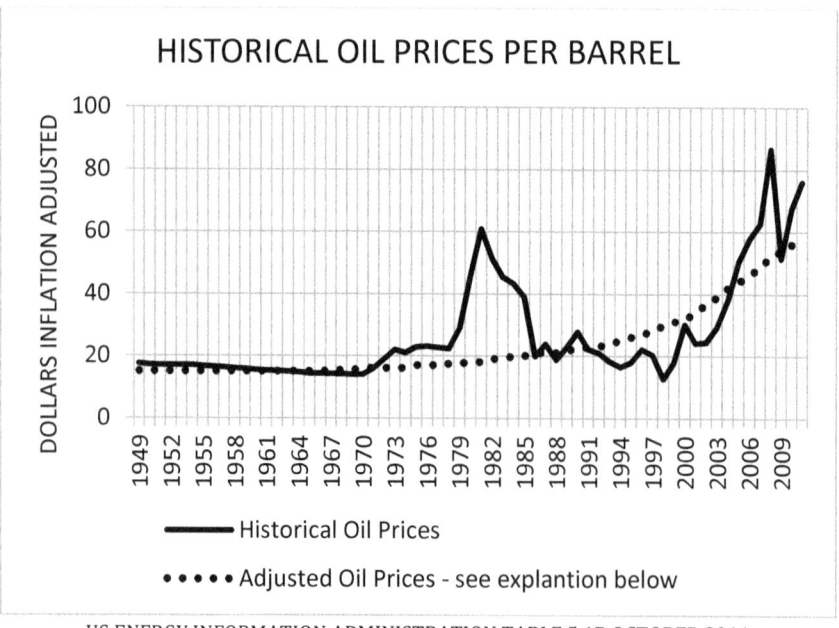

US ENERGY INFORMATION ADMINISTRATION TABLE 5.1B OCTOBER 2011

Governments are aware of the destabilization of oil prices resulting from dependence on foreign governments, but they are not capable of dealing with the problem.

Political unrest in oil producing countries, and interference with the supply caused by hostile governments is the greatest cause of unstable oil prices. In terms of inflation adjusted dollars, oil prices remained stable from 1880 until the 1973 Yom Kippur War. After that there were numerous large fluctuations in oil prices due to political and economic unrest.

Speculation in oil markets by financial institutions and price manipulation by the OPEC producers can result in oil prices that do not follow the rule of supply and demand. For short periods of time such prices may not reflect the actual cost of producing oil. However, in the long term prices will reflect the actual cost of producing oil which in terms of inflation adjusted dollars will show a steady increase.

Strategic Petroleum Reserves (SPR) were developed as an emergency fuel storage of oil maintained by the United States Department of Energy. It is the largest emergency supply in the world with the capacity to hold up to 727 million barrels. However, the United States imports a net 12 million barrels of oil a day so the SPR holds about a 58-day supply. Strategic Petroleum Reserves have been used to stabilize oil prices with only short term results.

THE ENERGY FACTOR

We are not in the 18th century, when a loaf of bread was produced without the need for energy from fossil fuels. The energy to produce the wheat for that loaf of bread came from renewable manpower and horse power. Processing of the wheat for that loaf of bread did not consume any non-renewable fossil fuel. Let's give that 18th century loaf of bread an energy factor of zero (0) BTU.

Today, energy to produce the wheat for that loaf of bread is consumed by an air conditioned tractor with a large diesel engine powered by fossil energy. Processing of the wheat is done in an air conditioned automated factory running on electricity produced from non-renewable coal. That loaf of bread is shipped in a plastic bag which is made from our fossil reserves. If we add up the energy consumed by the tractor, the factory and the plastic bag, there will be a finite number that can be expressed in BTU's. The Department of Agriculture may be able to accurately estimate the number of BTU's consumed to manufacture that loaf of bread, but for this example, let's call the energy factor for that loaf of bread, 10,000 BTU. Using this same logic, we can place an energy factor on every product and service that we use.

We can further break down the energy factor for that loaf of bread into fossil energy and renewable energy. In the case of that loaf of bread, there is bad news and there is good news. The bad news is that diesel fuel used to power that tractor is obtained from fossil reserves which will eventually approach depletion and will be too expensive to power that tractor. The good news is that eventually that tractors and the delivery trucks will be powered by renewable bioethanol, Algae fuel, or batteries that can be recharged from electricity produced from renewable sources such as wind power. The plastic bag will be replaced by a bag made from renewable paper, and the factory will be powered by electricity from renewable wind power, or possibly from nuclear energy or geothermal energy.

Based on a relatively small energy factor for that loaf of bread, we can be assured of having bread available at an affordable price for everyone in the United States for many years into the future.

Driving a gas guzzling car to the market to pick up that loaf of bread will consume a half gallon of gasoline having an energy factor of 15,000 BTU's. However, if we drive an electric car to the market, it will consume only 4,000 BTU. If we ride a bicycle to the market or walk, it will consume less than 300 BTU.

Let's make all consumers aware of the energy factor for everything we consume and every trip we take so that we will be prepared for the eventuality when those BTU's will no longer be available in unlimited amounts the way we have been accustomed and rationing of energy will be our way of life.

My fellow Americans, how many non-renewable BTU's have you consumed today driving your gas guzzling car to the market to get a loaf of bread packaged in a plastic bag.

ENERGY EQUATION

The basic energy equation is quite simple. REQUIREMENTS minus AVAILABILITY equal SHORTFALL. Consider oil first because it will enter depletion prior to the depletion of natural gas or coal.

REQUIREMENTS for energy will increase as population increases and as technology gives each person new ways to consume energy. As we consume more energy, availability will not be sufficient to meet demand, resulting in a shortfall. Based on the economic rule of "supply and demand", cost of energy will increase, thus tending to stabilizing demand, but stabilizing demand will not be enough, demand must be reduced.

Reducing demand will not be a pleasant task. President Jimmy Carter said this on April 18, 1977 when he gave a speech on energy policy. The speech started out by saying "Tonight I want to have an unpleasant talk with you about a problem unprecedented in our history". The nation did not want to hear the word "unpleasant". Political analysts claim the speech was the reason Jimmy Carter was not reelected for a second term.

AVAILABILITY of energy from fossil fuels will decrease as we deplete our reserves of fossil fuels. The unknown facing us is whether renewable energy, energy conservation and replacements for oil will develop fast enough to balance our depletion of fossil fuels, particularly oil. Increases in the price of energy, and particularly oil in the first half of the 21st century shows that availability of energy is not meeting demand. We have a shortfall, and it is starting to show in the slowdown of our economy.

The equation of "Demand – Availability = Shortfall" will always balance. Our objective must be to keep the shortfall as small as possible so as not to cause hardship and economic recession. Obviously, this can be done by reducing demand and increasing availability.

SHORTFALL will occur because there will not be enough availability to fill the needs for some services. Unless we incur rationing of oil, essential requirements will suffer from the shortfall while non-essential requirements will consume oil by those corporations and individuals best able to afford and squander the limited available oil.

The looming energy crisis will be led by shortages of mobility fuels which are defined as the sources of energy used to propel moving vehicles such as road vehicles, airplanes and ships. The sources of these mobility fuels are dominantly derived from oil, with fuels such as electricity, renewable fuels, and nuclear energy lagging far behind. Oil production in the United States entered the depletion phase in 1970 when production started to decline. Oil production in the rest of the world will enter the depletion phase about 2020 when total word oil production starts to decline.

Shortfall could be reduced by using better insulation in our homes, mass transportation, hybrid cars and funny looking light bulbs. These efforts are being tried and have shown little effect on balancing the equation. Demand for energy could be reduced by rationing. However, the word "rationing" has the connotation that everyone would have less of whatever is being rationed. That will not be the case.

RATIONING

Rationing could be an equitable means of distributing a resource that is in short supply, if administered in an objective manner. Rationing could reduce squandering of fuel that is in short supply by not allowing fossil fuel to be used in non-essential "toys" such as boats, private planes, motor homes and other non-essential gas guzzlers. What will happen to all those toys that are already out there and what will happen to the industries that produce those toys? Those toys will be powered by more expensive renewable fuels such as Algae fuel. Purchasing the more expensive renewable fuels by consumers who can afford those expensive toys will stimulate the development and production of alternate fuels such as Algae, which will result in making them more readily available.

The only practical solution to reducing our shortfall is to make more renewable energy available. Rationing is one method of accelerating the development of economically practical renewable energy such as Algae fuel. It will be a difficult task to have consumers accepts rationing, but, if consumers can be convinced that we have an energy crisis and that rationing is a practical and equitable solution to that crisis, we will have taken a giant step to resolving our energy crisis.

Rationing will not come from our political or industrial leaders. The necessity for rationing will be realized as the result of demonstrations by our middle class consumers when they see themselves caught in a financial depression caused by shortages of oil, and at the same time see the wealthy squandering oil on their gas guzzling toys.

Some people may remember World War II when it was necessary to ration gasoline. Consumers realized the need for rationing and there was very little objection. The system at that time for rationing in terms of 21st century technology was primitive. Now with modern electronic technology we can have a rationing system that will be equitable, accurate and honest. As time and technology provide us with new renewable fuels such as Algae fuels, rationing could be easily adjusted to meet the changing markets.

Rationing based on electronic technology will have a secondary beneficial effect by providing our regulators with accurate data on how fuel is being distributed and how fuel is being squandered.

If the government of the United States takes drastic action such as rationing of depleting resources, future starvation and suffering can be prevented. Will our government take action before it is too late? PROBABLY NOT

WHO IS AT THE CONTROLS?

There are two expenses that consume the major portion of low and middle class consumer's wealth. First is the mortgage on their home. Second is their payment for energy in all forms from the gasoline they put into their cars to the energy that is required to produce the milk that they drink every day.

First let's look at the housing mortgage industry where large financial firms made bad investments in high risk mortgages, misinformed their investors of the risks, and paid credit rating agencies to give false credit ratings to the mortgages so that they could resell them to their clients who trusted them and did not have the necessary information to properly evaluate their investments. This greed eventually resulted in the financial crisis of 2008 which will affect the world for many years after. The greed paid off to a few at the expense of many millions who lost their homes and jobs.

Our leaders were informed on many occasions that we were on a coarse to disaster, and no action was taken.

Second, let's look at the greed of the large oil companies who are making enormous profits from oil depletion. They will not admit that United States peak oil occurred in 1970 and that world peak oil is starting to occur at the present time. Instead, they are trying to sell their other product; natural gas which they claim is safe and plentiful. This book proves otherwise.

There is an analogy between the 2008 financial crisis caused by the housing bubble bursting and the 2011 world financial crisis which is a combination of the continuing 2008 financial crisis and the energy shortfall adding to unemployment and lack of economic growth.

The above analogy shows the lack of political incentive to institute regulation and oversight to prevent a looming disaster.

DEBT, DEFICIT AND BANKRUPTCY

Trade deficit is the most significant barrier to economic growth and jobs creation. Its effects are more formidable than the federal budget deficit.

Trade deficit is said to be self-correcting. As an example, if we buy large amounts of goods from China but in turn, they do not buy an equal value of good from us, we will have a trade deficit with China. That trade deficit will cause the value of the U.S. dollar to decrease compared to the Chinese Yuan which will give reason for the Chinese to buy more goods from the U.S. However, when the value of the dollar decreases, the U.S. will effectively be paying more for oil and imported materials that are necessary to produce goods for internal consumption as well as for exports.

Here is the connection between debt and deficit: Resulting higher prices for U.S. goods resulting from imported materials, particularly oil, will cause less demand, less production of goods and less need for employment of production workers. The financial burden caused by unemployment cannot be made up by increasing taxes, so the government must borrow money by issuing treasury notes which increase the national debt. Interest paid on the treasury notes further increases the national debt.

Both trade deficit and national debt can result in bankruptcy. If a private company cannot pay for the materials necessary to produce their products, and they can no longer obtain credit because they are considered a bad risk, their suppliers of materials will stop deliveries. Without materials, the company will go bankrupt and cease operations. However, if the government considers the corporation or financial institution essential to the economy of the nation, the government will bail them out by loaning them money or by outright giving them money. Where does the government get the money? It issues treasury notes which it must repay with interest. These notes further increase the national debt to the level where the government may not even be able to pay the interest unless it cuts back on essential services. Even if it does cut back on essential services, the treasury notes will be considered a greater risk and therefore will carry a greater interest burden. This starts a vicious cycle which could result in bankruptcy. Could a nation go bankrupt? Yes. However, if is a small nation such as Greece goes bankrupt, the international community will come to the rescue. What happens if it is the United States goes bankrupt?

Here is how depletion of oil reserves will cause bankruptcy. We have already seen how depletion of world oil reserves cause increased prices for gasoline at the pump. As gasoline prices increase, all goods and services using gasoline will increase in cost. That means the price for all goods and services will increase. Consumers will not be able to purchase both essential and non-essential goods and services, resulting in unemployment, which will reduce federal income from employment taxes. The government will take several courses of action. First, they will try to hold down the price of gasoline at the pump by releasing oil from the strategic petroleum reserves. When oil is purchased to replenish the strategic petroleum reserves, the price of gasoline will be even higher than before. Second, the government will try to stabilize the price of gasoline with direct subsidies which will add to the national debt. Third, the government will start to subsidize production of renewable energy such as algae bio-diesel which will be too little and too late and will cause a further increase in the national debt with resulting consequences.

In order to put the U.S. trade deficit, and the corresponding part of the deficit resulting from oil imports into perspective, the following graph shows a sharp increase in oil imports corresponding with total trade deficit starting about 1998.

The graph shows a decline in both trade deficit and U.S. oil imports from 2007 to 2009 which is attributed to the mortgage backed securities market bubble bursting.

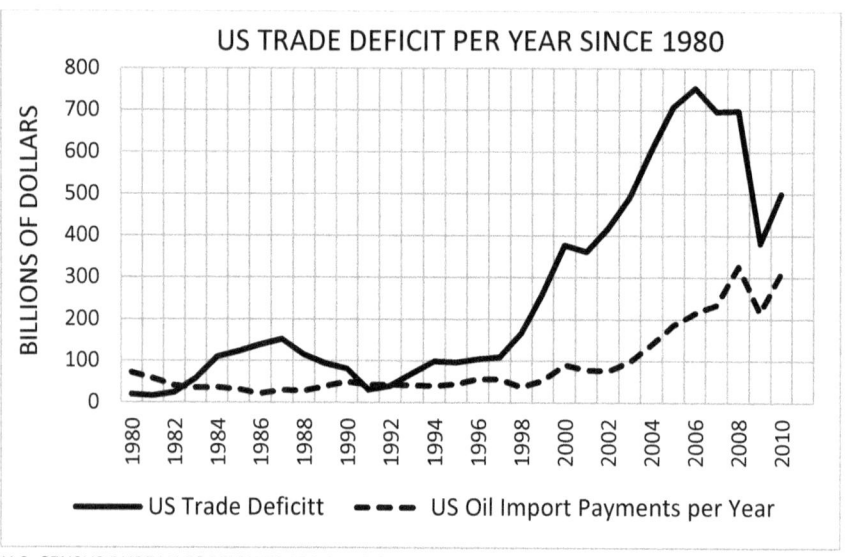

U.S. CENSUS BUREAU FOREIGN TRADE DIVISION

Referring to the above chart titled "US Trade Deficit per Year Since 1980", oil imports in 2010 accounted for over 60% of the US trade deficit and is increasing at an alarming rate. The dip in trade deficit in 2008 was caused by the 2008 Great Recession.

Abrupt reductions in Trade Deficit and Oil Import Payments between 2008 and 2009 can be attributed to the Great Recession of 2008. What went wrong to cause the great recession? Some say the recession was caused by the housing mortgage meltdown. All say that it was caused by a breakdown in the US government lack of regulation and oversight of the securities and banking industries.

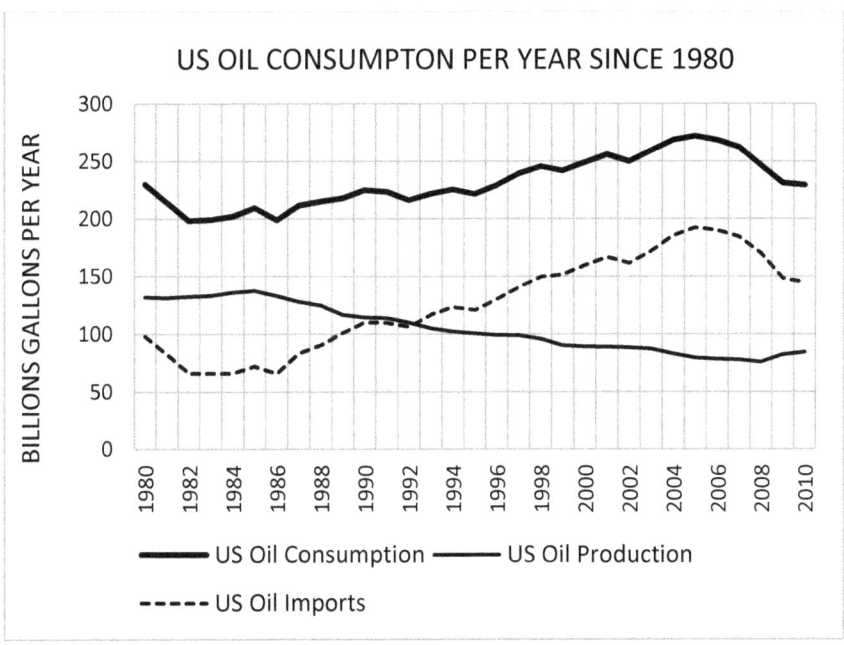

US OIL CONSUMPTON PER YEAR SINCE 1980

U.S. ENERGY INFORMATION ADMINISTRATION
INDEPENDENT STATISTICS AND ANALYSIS

Referring to the chart titled "US Oil Imports per Year Since 1980", US oil consumption and imports have been increasing while US oil production has been decreasing.

The trend of decreasing US oil production may be temporarily reversed as the result of recent developments of shale oil from areas such as the Bakken oil fields in Montana, North Dakota, and Saskatchewan

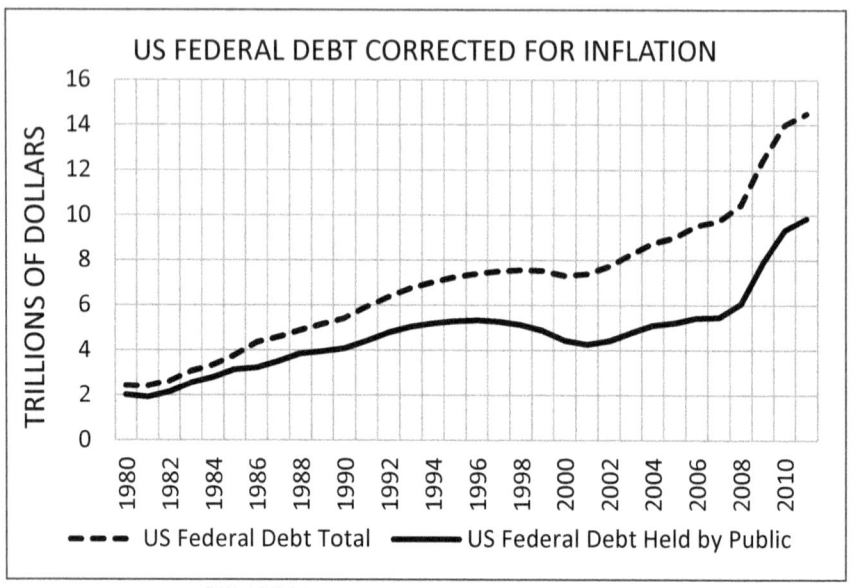

BUDGET OF THE UNITED STATES GOVERNMENT
FISCAL YEAR 2008 HISTORICAL TABLE 7.1

US FEDERAL DEBT TOTAL

The US Federal debt (dash line) consists of all indebtedness of the treasury of the United States government. As of February 2011 the total federal debt was 14.2 trillion dollars. The total or gross federal debt is the sum of the "debt held by the public" and "intragovernmental" debt.

US FEDERAL DEBT HELD BY THE PUBLIC

Public debt (solid line) represents all federal securities held by institutions or individuals outside the United States government, usually in the form of Treasury Notes. As of February 2011, total marketable securities were $9.6 trillion dollars.

INTRAGOVERNMENTAL HOLDINGS

Difference between the dash and solid lines represents US Treasury securities held in accounts which are administered by the United States government, such as Social Security and Medicare.

DEPENDENCE ON FOREIGN SUPPLIERS

The following paragraphs draw an analogy between the dependence of a small candy store for their supply of chocolate, with the dependence of the United States for their supply of oil.

Cathy's Candy Corner is a store owned and operated by the Collins family in Candyville USA, manufacturing and selling candy. Some of the ingredients in the candy included sugar, milk, fruit, and chocolate. All the ingredients except for the chocolate were purchased locally from suppliers they referred to as "family". Because of weather conditions, the chocolate could not be grown locally so they had to purchase it from Peters Peculiar People who became envious of the success of the Collins family but grudgingly supplied them with chocolate because they wanted the cash. One day Peter's Peculiar People decided to stop supplying chocolate to the Collins family. Because of the "embargo" of chocolate, Cathy's Candy Corner could not make enough candy for all the people who had become addicted to chocolate covered candy. There were long lines of people waiting to get candy. Meanwhile on the other side of town:

Donna's Delicious Delights continued to purchase chocolate from Peter's Peculiar People but had to pay more and more for their chocolate which they passed on to their customers. Business was doing great until their price for candy reached a point where people could no longer afford the chocolate candies. Sales declined and some of their employees were put out of work. With fewer sales, Donna's Delicious Delights had to borrow money to stay in business. Now, they not only had the burden of more expensive chocolate, but also had to pay interest on the money they borrowed to pay for the chocolate, and so, eventually they went bankrupt. Their employees were out of work and could not afford to purchase goods from neighboring businesses. Other businesses were adversely affected and had to lay off employees. And so the chocolate bubble started to burst.

As the bubble started to burst, there was less revenue for maintaining essential services in Candyville USA. The town leaders had to make a decision between raising taxes and borrowing money. Raising taxes has always been a bad thing politically, so in order to stay on good terms with the people who elected them, the politicians decided to borrow more money, which only temporarily delayed increasing taxes, until debt service on the new borrowed money had to be added to the tax rate. The chocolate bubble continued to burst, all because Donna's Delicious Delights continued to buy chocolate from Peter's Peculiar People rather than change the way they made candy. While Donna's Delicious Delights was borrowing money to stay in business:

Cathy's Candy Corner decided to make candy with less chocolate and more fruit. Their new candies were more nutritious and healthier, but people were reluctant to make the change. The chocoholics wanted the immediate satisfaction from chocolate that they had become addicted. Slowly, sales of the fruit flavored candies increased and Cathy's Candy Corner was pleased to have their entire business "in the family" without having to worry about the whims of Peter's Peculiar People. The chocoholics eventually agreed that the new candies were better for their health and for their "family". The economy of Candyville USA improved and the chocoholics changed their addiction from chocolate to fruit candies.

A not so simple example of debt service is a nation (the United States) with depleting reserves of essential oil. In order to maintain the extravagant life styles of the voting public, the government borrows money to purchase essential oil from foreign countries. In order to pay the debt service on that borrowing, it is necessary for the government to burden its population with higher taxes. With higher taxes, consumers have less to spend on the goods and services that drive the economy. Thus, because of the ever increasing debt service the economy falters causing suffering and starvation.

GROSS DOMESTIC PRODUCT (GDP) VS. ENERGY CONSUMPTION 2007

Country	GDP per capita	Million BTU per capita
Norway	78,436	232.0
Canada	59,599	328.4
Denmark	55,870	143.3
Finland	46,505	249.2
United Kingdom	46,096	137.1
United States	45,744	308.4
France	40,460	169.6
Germany	40,468	161.1
Singapore	38,465	169.0
Japan	34,264	160.5
New Zealand	32,712	157.1
Greece	27,767	107.5
South Korea	21,653	182.5
Hungry	13,713	105.8
Mexico	9,848	66.6
Turkey	8,865	54.5
Venezuela	8,243	92.1
Argentina	7,624	74.1
Costa Rica	6,385	44.5
South Africa	5,930	110.8
China	3,748	59.3
Egypt	2,276	33.4
Sri Lanka	2,072	18.4
India	1,195	21.1
Pakistan	954	20.6
Vietnam	835	26.3
Haiti	645	11.4
Bangladesh	550	6.7

DATA FROM THE WORLD BANK IS FOR YEARS PRIOR TO 2007 SO AS TO REDUCE DISTORTION DUE TO THE 2008 RECESSION.

The previous table was made in an attempt to determine why nations spend different percentages of their GDP per capita on energy.

Average energy consumption per capita worldwide for 2007 was 70 million BTU (Million BTU per Capita). The United States and Canada had 4½ times the world average energy consumption per capita.

United States and Canada consume large amounts of energy, mostly in the transportation sector, and have low fuel prices compared with Europe and Asia.

Norway has a high GDP per capita because it is a petroleum exporting country.

South Africa consumes a large amount of energy per capita on energy because of its large scale energy intensive mining industry which require large amounts of energy.

The wealthier nation's consumption of energy per capita in BTU's is a lesser percentage of their income than the poorer nations. This can be attributed to the necessity that poor people consume a large percentage of their earnings on fuel for necessities such as cooking, washing, fuel, local transportation, etc.

THE CURSE OF TECHNOLOGY

Technology powered the industrial revolution which put more goods in the hands of more people creating more wealth which in turn put still more goods in the hands of still more people. Economists call this prosperity. The goods all had a common denominator, which was they all required energy to produce. In addition to consuming energy to produce, many of the goods such as automobiles consumed energy for their operation. At the beginning of the industrial revolution, little thought was given to how much of our fossil reserves were being consumed and how long our fossil reserves would last.

In 1973, the United States was awakened to the fact we were dependent upon foreign nations to supply us with enough oil to sustain our wonderful industrial revolution. Much rhetoric was given to have the United States become energy independent but it turned out to be merely rhetoric.

Politicians and consumers believed that technology got us into this situation and technology would solve our problems. Yes, new sources of energy and new methods of conservation are being developed, but they are coming too little and too late. Compulsory conservation of energy by extreme measures such as rationing may become necessary if technology does not get us out of this mess real soon.

We must conclude that the wealth of the United States results in squandering its natural resources, particularly oil, which is causing it to become more dependent on foreign oil, which eventually could cause its economic collapse.

WORLD WEALTH

Statistics show that the separation between wealthy and poor is constantly increasing. Combine this with the fact that never in the history of mankind has there been an occasion when an essential resource such as oil has entered its depletion phase. Our political and industrial leaders do not know how to deal with such a situation.

If we do not have equitable distribution of our depleting energy reserves, the wealthy will squander our nation's reserves on their gas guzzling vehicles, yachts and jet airplanes, while the poor will not have enough energy to heat their homes or drive to work. What will happen? Let's not find out, because things could get very ugly.

Distribution of wealth throughout the world can be understood by some simple facts. The richest 1% of people in the world own 40% of the wealth. Those in financial services and the internet sectors predominate among the super-rich. Europe, the US and some Asia Pacific nations account for most of the extremely wealthy. US accounts for 33% of the total, Japan accounts for 27% of the total, the UK for 6% and France for 5%.

Per capita wealth of the three wealthiest nations is Japanese having $181,000, Americans having $144,000, and UK residents have on average $127,000.

World Institute for Development Economics Research of the United Nations - is the first to chart wealth distribution in every country as opposed to just income. It included all the most significant components of household wealth, including financial assets and debts, land, buildings and other tangible property. Together these total $125 trillion globally.

The study affirmed the existence of a nest egg to provide an insurance policy to help people cope with unforeseen events such as ill health or loss of employment. Capital allowed people to drag themselves out of poverty. It was more difficult in developing countries to set up a business because it was harder to borrow start-up funds.

The study found the richest 10% of adults accounted for 85% of the world total of global assets. Half the world's adult population, however, owned barely 1% of global wealth. Near the bottom of the list was India, with per capita wealth of $1,100, and Indonesia with $1,400.

Many African nations as well as North Korea and the poorer Asia Pacific nations were places where the worst off lived.

WHAT CAN BE DONE?

United States economists and politicians seem to agree (wrongly) that reducing income taxes will stimulate the economy, and that the increase in debt caused by reduction of tax revenue will be temporary until the positive effects of the stimulated economy are seen as increased tax revenues. Reducing taxes in 2010 and 2011 did not have the anticipated effect because of consumer lack of confidence in the economy. Instead of spending the money saved from lower taxes, consumers resorted to saving the money until the economy improved.

Economists and politicians agree (wrongly) that throwing money at businesses will increase employment which will in turn provide additional tax revenues that will reduce the temporary increase in the debt. This type of stimulus was tried in 2010 and 2011 by the Obama administration with very little positive effect.

Reducing taxes and providing more money for businesses will stimulate the economy only if that money is spent, not squirreled away. The United States needs a revision in its tax structure that will encourage spending on service items most of which do not consume large amounts of fossil fuels.

The United States needs a federal tax system that will stimulate the economy, reduce the widening gap of wealth between upper class and lower class citizens, and reduce our extravagant use of fossil fuels. Is such a tax practical? Answer:

YES. *Are our politicians willing and able to legislate such taxes? Answer: NO.*

How about this? At present, every family that owns a home pays a personal property tax (tangible property tax) on the home, better known as the real estate tax. Before buying a house they will always ask about the real estate tax, because it part of the cost of ownership. There are few complaints about real estate taxes because most consumers consider it to be an equitable tax. Some states such as South Carolina have taken the personal property tax a step further to cover personal property such as automobiles, boats, and other big ticket items. The personal property tax is an equitable tax that discourages gas guzzling properties in favor of service spending which employs workers in less energy intensive professions. On the State level, the personal property tax can be readily enforced.

On the federal level, a personal property tax would be a political nightmare with the party representing the middle class in favor and the party representing the wealthy opposed to a personal property tax. If properly administered, a personal property tax at the federal level would stimulate the economy and would be readily accepted as an equitable tax.

Florida had an Intangible Property Tax until 2007 when Florida Governor Jeb Bush repealed the tax. Prior to its repeal, it was considered an equitable tax favoring the middle class. Common types of property covered by the tax were: stocks, corporate and government bonds issued by government entities outside of Florida, and shares or other types of ownership units in mutual or money market funds. For lack of a better explanation, repeal of the Florida Intangible Property Tax was the result of influence by large corporations and wealthy residents. This is another example of the wealthy politicians controlling taxes in favor of the wealthy.

Tangible and intangible property taxes could be referred to as "Use it or Lose it Taxes", which will have the dual purpose of stimulating the economy and at the same time reducing the national debt.

TAX ON DEPLETING RESOURCES

The United States is consuming more oil from combined internal production and imports per capita than any other country. In 1970, production of oil in the United States peaked which resulted in increased dependency upon foreign oil to meet demands. This dependency on foreign oil is not being reduced in spite of all efforts to reduce our consumption of oil by growing Ethanol, driving Hybrid cars, constructing Windmill Farms and using funny looking light bulbs.

United States demand for oil must be reduced. In Europe and other foreign countries, increasing the tax on gasoline has resulted in smaller more efficient cars. In the United States, where low oil prices are our way of life, increasing the tax on gasoline or home heating oil will be unacceptable having political repercussions.

An alternative to increasing the tax on gasoline, would be to place a tax on oil as it enters the US, whether it comes from below the ground in the United States or a pipeline from Canada or an oil tanker from Saudi Arabia. Such an entry tax would discourage extravagant usage of oil products without political repercussions. If administered prudently, the entry tax could be increased slowly and predictably so as to give consumers time to plan their next purchase of a car or improve the insulation of their home, or consider the feasibility of solar panels on the roof of their home.

INVESTMENTS IN ENERGY INFRASTRUCTURES

As fossil fuels, particularly oil, become depleted, liquid mobility fuels for land based vehicles will be Algae fuel, Ethanol, biodiesel, or liquid hydrogen. Competing with liquid mobility fuels will be electric vehicles powered from self-contained batteries or electric trolley wires.

Technology and electricity costs will determine which mobility fuels will dominate our future transportation. It will be many years until our free market system determines which form of liquid mobility fuel dominates transportation. The development costs of liquid mobility fuels can be undertaken by industry without government intervention. However;

The switch to electricity in land based transportation has already started with more and more electric cars, trucks, busses and trains being put into service. These vehicles can be classified as either trolley wired, short fixed route commercial or private short commute vehicles

Overhead trolley wires for electric vehicles such as long haul trains will be too great an investment for industry to finance. Therefore, overhead trolley wires and battery charging stations would be a good federal investment in infrastructure which at the same time could help alleviate our unemployment problems.

The United States became great because it understood the benefits of investing in the future. Investments were made in hydroelectric power, railroads, highways, and aviation. Today they are still some of our greatest assets.

The clock is ticking on oil depletion. Now is the time to start planning for electric transportation.

Exceptions to electric transportation are the airline and water transport industries. Small airplanes and small boats have been propelled by batteries, providing clean inexpensive transportation. However, large scale use of electricity for commercial planes and ships will not be practical in foreseeable future. Nuclear propulsion for large ships and alternative fuels for commercial aircraft are discussed in Section Eight, Future of Energy.

SECTION FIVE
SOURCES OF ENERGY

MAN, horses, wind and water power provided all the energy needs of mankind prior to the 18th century. This energy was 99.9% renewable. This goes back to the beginning of mankind when human muscle was used to build the Pyramids and the Great Wall of China. Workers in those days had no idea that in the distant future, a single machine would replace the efforts of many workers, nor could they have realized that machines would get their power from that messy oil that was oozing out of the ground. Scientists had no idea that oil would be taken out of the ground to power machines, and that it would be depleted within the next 500 years and that oil could not be renewed by nature for the nest several million years.

Everything was renewable and there was no reason to be concerned that man would damage the environment. There was no such thing as plastic that would pollute the environment for hundreds of years. Wars were being fought so that one nation could control another nation or possibly the world. There was no thought that wars of the future would be fought over that messy oil oozing out of the ground.

Here we are in the 21st century and machines have replaced manual labor in many industries. Creation of jobs in the 21st century is for office workers sitting in front of a computer for 8 hours. When they get home from work, they sit in front of a TV screen and eat too much of the wrong kind of food. Result is that the population is becoming over weight and in poor health. Answer is the exercise machine that replaces the need for healthy manual labor. Is technology providing us with a better and healthier life?

HORSES were domesticated to assist man to perform many of his everyday tasks and for transportation. Horses were kind to each other and to man. They were powered by renewable energy and didn't pollute the environment. In the 21st century, horses are still used on many farms and for transportation. Horses are presently being used by the Amish for farming and transportation. The Amish use less energy per person than any other group in the United States. Do the Amish know something to which the rest of the world should be paying attention?

WIND POWER was used by sailing ships in ancient Egyptian time. Their ships were about 80 feet in length and had a single mast. These ships were probably the first examples of man utilizing readily available renewable energy for transportation. Utilization of wind to power ships continued to develop so that in 1492, the famous voyage of Christopher Columbus was possible. Wind powered sailing ships continued to be the primary means of transportation until the development of steam engines which replaced wind power. In modern times, wind powered sailing ships are a popular form of recreational transportation.

The great clipper ships that sailed the world prior to the 19th century were constructed from renewable materials, used renewable energy, and did not pollute the environment.

There was a negative side to the great clipper ships. They were not fast enough for man to accomplish his objectives of consuming the natural resources of the world and polluting the environment, so man developed faster ships that consume our natural resources burning coal and in later years even faster ships that burn oil.

WIND MILLS came into being before 1274, driven by rivers and streams. They had become a familiar sight in 1414. The earliest known drainage mills were invented and by 1450 many could be found in South Holland. The development of Wind Mills is most certainly attributable to Holland as it is unrivaled in the diversity of types of mills. The advent of technology, however, brought a quick end to the mill's usefulness. First the steam engines, the internal combustion and finally the electric motor, all gradually took over the jobs previously undertaken by the wind or the water. At the start of the 21st century, only 1000 remain. Due to imminent depletion of oil reserves in the 20th century, modern Wind Mills (Wind Turbines) are coming back into popularity as a primary producer of renewable electric energy.

WATERPOWER increased dramatically in 1793 when Samuel Slater built the country's first successful water powered textile mill on the Blackstone River in Pawtucket, Rhode Island. Slater's success formed the basis for a new economy in New England - one much more lucrative than grinding grain or sawing logs. Small gristmills and sawmills throughout New England were replaced with water-powered textile mills. In the 21st century water power is no longer used to drive textile mills which are now powered by electricity obtained from large electric generators, some of which are hydroelectric. However, small hydroelectric generators are being constructed in developing nations to harness power from small streams and lakes.

HYDROELECTRICITY is the term referring to electricity generated by hydropower; the production of electrical power through the use of gravitational force of falling or flowing water. It is the most widely used form of renewable energy. Once a hydroelectric complex is constructed, the project produces no direct waste, and has a considerably power output level of the greenhouse gas carbon dioxide (CO_2) than fossil fuel powered energy plants.

Hydropower is the cheapest way to generate electricity today. That's because once a dam has been built and the equipment installed, the energy source—flowing water—is free. It's a clean fuel source that is renewable yearly by snow and rainfall. Hydropower is readily available when consumer demand requires more or less electricity. Engineers control the flow of water through the turbines to produce electricity on demand by simply opening or closing flow control valves.

In 1878 the world's first hydroelectric power scheme was developed at Cragside in Northumberland, England by William George Armstrong. It was used to power a single light bulb in his art gallery. The old Schoelkopf Power Station No. 1 near Niagara Falls in the U.S. side began to produce electricity in 1881. The first Edison hydroelectric power plant, the Vulcan Street Plant, began operating September 30, 1882, in Appleton, Wisconsin, with an output of about 12.5 kilowatts. By 1886 there were 45 hydroelectric power plants in the U.S. and Canada. By 1889 there were 200 in the U.S. alone. Hydroelectricity in the United States was made famous by the Hoover Dam in Nevada which has installed capacity of 2080 MWe megawatts of electricity equivalent to two nuclear power stations.

United States has 8,100 major hydroelectric dams producing 79,500 MWe megawatts of electricity, (2.38 quadrillion BTU per year) equivalent to approximately 80 nuclear power stations.

Worldwide hydroelectricity installed capacity is 777,000 MWe megawatts of electricity, (23.22 quadrillion BTU per year). This was approximately 20% of the world's electricity, and about 88% of electricity from renewable sources. Tabulation and graphs of hydroelectricity are shown in Section Eight, Energy Statistics.

GEOTHERMAL RESOURCES are theoretically more than adequate to supply humanity's energy needs, but only a very small fraction may be profitably exploited. Drilling and exploration for deep resources is very expensive. Forecasts for the future of geothermal power depend on assumptions about technology, energy prices, subsidies, and interest rates. Geothermal power is considered to be sustainable because any projected heat extraction is small compared to the Earth's heat content.

Even though geothermal power is globally sustainable, extraction must still be monitored to avoid local depletion. Over the course of decades, individual wells draw down local temperatures and water levels until a new equilibrium is reached with natural flows. The three oldest sites, at Larderello, Wairakei, and the Geysers have experienced reduced output because of local depletion. Heat and water, in uncertain proportions, were extracted faster than they were replenished.

The long-term sustainability of geothermal energy has been demonstrated at the Lardarello field in Italy since 1913, at the Wairakei field in New Zealand since 1958, and at The Geysers field in California since 1960.

The Geysers in the Mayacamas Mountains, located north of San Francisco, naturally occurring steam field reservoirs below the earth's surface are being harnessed by Calpine to make clean, green, renewable energy for homes and businesses across Northern California.

The Geysers, comprising 45 square miles along the Sonoma and Lake County border, is the largest complex of geothermal power plants in the world. Calpine, the largest geothermal power producer in the U.S., owns and operates 15 power plants at The Geysers with a net generating capacity of about 725 megawatts of electricity - enough to power 725,000 homes, or a city the size of San Francisco.

The Geysers meets the typical power needs of Sonoma, Lake, and Mendocino counties, as well a portion of the power needs of

Marin and Napa counties. In fact, The Geysers satisfies nearly 60 percent of the average electricity demand in the North Coast region from the Golden Gate Bridge to the Oregon border. The Geysers is one of the most reliable energy sources in California delivering extremely high availability and on-line performance and accounts for one-fifth of the green power produced in California.

ETHANOL FUEL (ethyl alcohol), the same type of alcohol found in alcoholic beverages. It is most often used as a motor fuel, mainly as a biofuel additive for gasoline. World ethanol production for transport fuel tripled between 2000 and 2007 from 4.5 billion to more than 14 billion gallons. From 2007 to 2008, the share of ethanol in global gasoline type fuel use increased from 3.7% to 5.4%. In 2009 worldwide ethanol fuel production reached 19.5 billion gallons.

Ethanol fuel is widely used in Brazil and in the United States, and together both countries were responsible for 86 percent of the world's ethanol fuel production in 2009. Most cars on the road today in the U.S. can run on blends of up to 10% ethanol, and the use of 10% ethanol gasoline is mandated in some U.S. states and cities. Since 1976 the Brazilian government has made it mandatory to blend ethanol with gasoline, and since 2007 the legal blend is around 25% ethanol and 75% gasoline. In addition, by December 2010 Brazil had a fleet of 12 million flex-fuel automobiles and light trucks and over 500 thousand flex-fuel motorcycles regularly using neat ethanol fuel.

Bioethanol is a form of renewable energy that can be produced from agricultural feedstock. It can be made from very common crops such as sugar cane, potato, manioc and corn. However, there has been considerable debate about how useful bioethanol will be in replacing gasoline. Concerns about its production and use relate to increased food prices due to the large amount of arable land required for crops, as well as the energy and pollution balance of the whole cycle of ethanol production, especially from corn. Recent developments with cellulosic ethanol production and com-mercialization may allay some of these concerns.

Cellulosic ethanol offers promise because cellulose fibers, a major and universal component in plant cells walls, can be used to produce ethanol. According to the International Energy Agency, cellulosic ethanol could allow ethanol fuels to play a much bigger role in the future than previously thought.

METHANOL is an alternative fuel for internal combustion engines, either in combination with gasoline in automobiles or directly used in racing cars. Methanol has received less attention than ethanol fuel, because in the 2000s particularly, the support of corn-based ethanol offered certain political advantages. In general, ethanol is less toxic and has higher energy density, although methanol is less expensive to produce and is a less expensive way to reduce the carbon footprint. However, for optimizing engine performance, fuel availability, toxicity and political advantage, a blend of ethanol, methanol and petroleum is likely to be preferable to using any of these individual substances alone. Methanol may be made from fossil or renewable resources, in particular natural gas and biomass respectively.

Methanol can be produced by the destructive distillation of wood, resulting in its common name of wood alcohol. Methanol can be produced using methane from natural gas as a raw material. In China, methanol is made for fuel from coal.

BIODIESEL refers to a vegetable oil- or animal fat-based diesel fuel consisting of long-chain alkyl (methyl, propyl or ethyl) esters. Biodiesel is meant to be used in standard diesel engines and is thus distinct from the vegetable and waste oils used to fuel *converted* diesel engines. Biodiesel can be used alone, or blended with petrodiesel. Biodiesel can also be used as a low carbon alternative to heating oil.

Documentary film "Fuel" is based on biodiesel, made from vegetable oil, as a viable alternative to fossil fuels. Although politicians and energy execs have done their best to quell it, the benefits of biodiesel are real. This documentary (winner of the Grand Jury Prize at Sundance) chronicles the quest to popularize the untraditional fuel source, citing the environmental and economic advantages the country could reap by adopting it.

ALGAE FUELS - February 2010, the Defense Advanced Research Projects Agency announced that the U.S. military was about to begin large-scale production of oil from algal ponds into jet fuel. November 2011, the first commercial flight using algae biofuel flew from Houston, Texas to Chicago O'Hare airport. Further discussion of algae fuels is included in Section Eight, Future of Energy. Are we just starting to see the tip of the iceberg when it comes to the potential of algae fuel.

SOLID BIOMASS is material of recent biological origin. The list of solid biomass sources includes plant matter and other biodegradable wastes. It is used primarily as a combustible fuel. Biomass may be used as a substitute for standard fossil fuels. It is an environment friendly and it is a sustainable resource. The most common form of solid biomass is wood fuel. Unused crops from fields are also used to generate energy. Municipal solid waste is also used as a feed source. Some crops are planted and grown expressly for the purpose. Cow dung is also included in the solid biomass list. Solid biomass is used to generate power and heat. A common example is the burning of wood. The internal dwelling environment is also heated by burning wood in colder countries. Other organic materials suitable for burning include wheat chaff, corn cobs, and sugar cane residue. Wood burning power plants are environmentally friendly because they use renewable fuel, have low emissions and the residual product can be used for productive purposes.

In terms of capacity, biomass power plants represent the second largest amount of renewable energy in the nation. Biomass power plants currently represent 11,000 MW - the second largest amount of renewable energy in the nation

Burlington, Vermont built the Joseph C. McNeil generating station which has been in operation for almost 30 years. It burns wood chips available locally. As a backup, the plant can use natural gas. The plant can produce 50 megawatts of electricity which is about one tenth the power from the Vermont Yankee nuclear power station. It is one tenth the physical size of Vermont Yankee and the cost of construction and maintenance is a small fraction of any nuclear power plant when compared on a dollar per watt basis. Compared to nuclear reactors, physical safety has never been an issue.

The Biomass Power Association is working to expand and advance the use of clean, renewable biomass power. It represents 80 biomass power plants in 20 states across the U.S.

MODERN WIND TURBINES used in wind farms for commercial production of electric power are usually three-bladed and pointed into the wind by computer-controlled motors. These have high tip speeds of over 320 kilometers per hour (200 mph), high efficiency, and low torque ripple, which contribute to good reliability. The blades are usually colored light gray to blend in with the clouds and range in length from 20 to 40 meters (66 to 130 ft) or more. The tubular steel towers range from 60 to 90 meters (200 to 300 ft) tall. The blades rotate at 10-22 revolutions per minute. At 22 rotations per minute the tip speed exceeds 300 feet per second. A gear box is commonly used for stepping up the speed of the generator, although designs may also use direct drive of an annular generator.

Because of their enormous size these modern wind turbines can be seen from many miles away. On a clear day, the pilot of a plane flying at 40,000 feet can count the rotations of the blades.

SOLAR PHOTOVOLTAIC is the method of generating electrical power by converting solar radiation into direct current electricity using semiconductors that exhibit the photovoltaic effect. Photovoltaic power generation employs solar panels composed of a number of cells containing a photovoltaic material. Materials presently used for photovoltaics include monocrystalline silicon, polycrystalline silicon, amorphous silicon, cadmium telluride, and copper indium selenide/sulfide. Due to the growing demand for renewable energy sources, the manufacturing of solar cells and photovoltaic arrays has advanced considerably in recent years.

As of 2010, solar photovoltaic panels generate electricity in more than 100 countries and, while yet comprising a tiny fraction of the 15 TW total global power-generating capacity from all sources, is the fastest growing power-generation technology in the world. Between 2004 and 2009, grid-connected PV capacity increased at an annual average rate of 60 percent, to some 21 GW. Such installations may be ground-mounted (and sometimes integrated with farming and grazing) or built into the roof or walls of a building, known as Driven by advances in technology and increases in manufacturing scale and sophistication, the cost of photovoltaics has declined steadily since the first solar cells were manufactured. Net metering and financial incentives, such as preferential feed-in tariffs for solar-generated electricity, have supported solar PV installations in many countries.

SOLAR THERMAL ENERGY This photo shows A rear view of some of the roughly 184,000 mirrors installed at Nevada Solar One, Boulder City, Nev. The mirrors direct sunlight on an oil-filled tube. The oil is then used to create steam, which turns a turbine. Solar Thermal Energy is a technology for harnessing solar energy for thermal energy (heat).

Solar thermal collectors are classified by the USA Energy Information Administration as low-, medium-, or high-temperature collectors. Low temperature collectors are flat plates generally used to heat swimming pools. Medium-temperature collectors are also usually flat plates but are used for heating water or air for residential and commercial use. High temperature collectors concentrate sunlight using mirrors or lenses and are generally used for electric power production. STE is different from and indeed much more efficient than photovoltaics, which convert solar energy directly into electricity. While only 600 megawatts of solar thermal power is up and running worldwide in October 2009 another 400 megawatts is under construction and there are 14,000 megawatts of the more serious concentrating solar thermal (CST) projects being developed.

Low-temperature collectors are generally installed to heat swimming pools, although they can also be used for space heating. Collectors can use air or water as the medium to transfer the heat to their destination.

COAL, a fossil fuel, is the largest source of energy for the generation of electricity worldwide, as well as one of the largest producers of carbon dioxide. Carbon dioxide produced from coal usage is slightly more than that from petroleum and about double the amount from natural gas.

Carbon dioxide is the largest source of greenhouse gases which is blamed for global warming. China produces 23% of the carbon dioxide in the atmosphere with the US producing 18%. Controversy over global warming was brought to the attention of the public by former Vice President Al Gore's book "An Inconvenient Truth". The controversy is ongoing.

Coal is primarily used as fuel to produce electricity and heat through combustion. World coal consumption is expected to increase 48% to 9.98 billion short tons by 2030. The US consumes about 14% of the world total, using 90% of it for generation of electricity.

COAL-FIRED POWER PLANTS (614) in the U.S. had a total of 1,522 coal-fired generating units and a total of 335,831 MW (Megawatts) of production capacity. Given total U.S. electric capacity of 1,075,677 MW, coal provided 31.2% of U.S. electric capacity in 2005 – second only to natural gas (41.2%) in the share of U.S. power capacity, and far ahead of nuclear (9.8%), hydroelectric (7.2%), oil (6.0%), or renewables (2.5%).

For many years - due to low cost and reliable production of coal, and the frequently unstable cost of natural gas - coal plants have been used more heavily than other power plants. In 2008, the U.S. produced approximately 2,133,000 GWh (gigawatt hours) of electricity from coal (one GWh is the amount of power produced by a 1,000 MW power plant running for one hour), accounting for about a quarter of the world's coal-fired electricity. For comparison in 2008 the largest producers of electrical power from coal were:

	2008 GWh	Quad BTU per year	% World Total
China	2,733,000	81,643	33.1
United States	2,123,000	63,455	25.8
India	569,000	17,007	6.9
Other Countries	2,838,000	162,148	34.2
World Total	8,263,000	246,974	

NATURAL GAS, as it comes from the ground is primarily Methane, a chemical compound which is the principal component of natural gas. As it comes from the ground, Natural Gas also contains heavier gaseous hydrocarbons: ethane , propane, butane, isobutene, pentanes and higher molecular weight hydrocarbons, acid gases, carbon dioxide, hydrogen sulfide These chemicals are collectively referred to as Natural Gas Liquids. Section Eight, Future of Energy, discusses the consequences of these Natural Gas Liquids when hydraulic fracturing is used to extract Gas from the ground.

NATURAL GAS FIRED POWER PLANTS. Because of its clean burning nature, natural gas has become a very popular fuel to generate electricity. In the 1970's and 80's, the choices for most electric utility generators were large coal or nuclear powered plants; but, due to economic, environmental, and technological changes, natural gas has become the fuel of choice for new power plants. In fact, in 2000, 23,453 MW (megawatts) of new electric capacity was added in the U.S. Of this, almost 95 percent, or 22,238 MW were natural gas fired additions. Natural gas fired electricity generation is expected to increase dramatically over the next 20 years, as all of the new capacity that is currently being constructed comes online. At the end of 2006 world gas-fired electric generation was 1,124 GWe of which 20% was from natural gas.

Natural gas fired electricity generation is expected to increase dramatically over the next 20 years, as new capacity that is currently being constructed comes online.

WHY OIL?

OIL is the largest source of energy in the world with consumption of 174 quadrillion BTU in 2007. Second is coal with consumption of 123 quadrillion BTU. A tabulation of all sources of energy is shown in Section Six, Energy Statistics

Oil and its derivatives can be used to power everything from transportation to manufacturing to electric generation. It is a convenient form of energy because it easily taken out of the earth. It is (or rather has been) a cheap form of energy because it takes little manpower to take it out of the earth. In addition to oil as a primary source of energy, many products used by man are based on oil such as plastics, medicines and lubricants.

As oil becomes more difficult to extract from the earth, it becomes more expensive. A land based oil rig is relatively simple compared to an offshore drilling rig. As land based oil becomes depleted and must be replaced by expensive offshore drilling rigs, the cost of oil will increase substantially as witnessed by increases in oil at the beginning of the 21st century. Because of this increase in the price of oil, there has been a shift to alternative energy sources such as biofuels, ethanol, wind and solar (nuclear is discussed separately). Those alternate sources of energy all require land mass to implement. In the case of ethanol, the land required to grow the feedstock results in an increase in the price of practically all foods, making it impractical as the long term solution for transportation fuel.

ON SHORE DRILLING RIGS. Oil has been known and used since the most ancient times and has been mentioned by most ancient historians since the time of Herodotus. It was used chiefly as a liniment or medicine, not as a fuel. Oil flows from natural springs in many localities. The first oil well was drilled in Pennsylvania by Edwin Drake in 1859. The well was 69 feet deep and produced 15 barrels a day. The area quickly boomed and the modern oil industry was born. Later Texas and Oklahoma became the centers of US production. Following World War II the Middle East became a major supplier to the US. As a fuel, oil was originally used as kerosene for lighting, replacing animal, vegetable and coal oils. It also came to be used in furnaces. Its biggest use, however, came with the development of the automobile. Today almost all forms of locomotion -- cars, trucks, buses, trains, ships and airplanes -- are fueled by oil, diesel or gasoline. Fuel oil has also been burned to produce electricity, although that has always been mostly coal's job.

OFFSHORE DRILLING typically refers to the discovery and development of oil and gas resources which lie underwater. Most commonly, the term is used to describe oil extraction off the coasts of continents, though the term can also apply to drilling in lakes and inland seas. Offshore drilling presents environmental challenges, especially in the Arctic or close to the shore. Controversies include the ongoing US offshore drilling debate. There are many different types of platforms for offshore drilling activities, from shallow-water steel jackets and jackup barges, to floating semisubmersibles and drillships able to operate in very deep waters.

OIL SANDS are located throughout the world with large deposits in Canada, Venezuela, USA, Russia and several other countries. Oil can be extracted from the sands by heat, chemical or mechanical processes depending on the temperature and viscosity of the oil. Oil sands reserves have only recently been considered to be part of the world's oil reserves, as higher oil prices and new technology enable them to be profitably extracted and upgraded to usable products. Because growth of oil sands production has exceeded declines in conventional crude oil production, Canada has become the largest supplier of oil and refined products to the United States, ahead of Saudi Arabia and Mexico.

Canada has oil sand reserves of 1.75 trillion barrels. About 173 billion barrels is estimated by the government of Alberta to be recoverable at current prices, using current technology

Canada's so-called "black gold" has come to be regarded as an abundant, secure, and affordable source of crude oil. But development of this unconventional fossil fuel comes with unconventional risks and consequences. Everything about the tar sands is big, most significantly its global warming and environmental implications, leading some to describe the tar sands as "Canada's dirty secret."

The United States needs the oil from Canada. The only practical means to transport oil from the Athabasca Oil Sands in northeastern Alberta, Canada to multiple destinations in the United States, which include refineries in Illinois, Oklahoma, with connections to refineries along the Gulf Coast of Texas is by the proposed Keystone Pipeline. Construction of the pipeline has drawn much political controversy in both the United States and Canada. As of early 2012, the Keystone Pipeline controversy has not been resolved.

 NUCLEAR ENERGY is the only source of energy that can provide the United States with enough electricity for at least the next 300 years. Reserves of Uranium beyond this are probably much greater but have not yet been proven. Reason for this is that the low cost of Uranium, does not warrant exploration to locate additional reserves. Uranium is said to be almost unlimited. For this reason, scientists consider nuclear energy to be in the category of a renewable source of energy.

The future of nuclear energy has become an emotional and political issue rather than a technical issue resulting from the negative publicity of nuclear accidents. There are three major issues that are keeping nuclear from eliminating our dependence on oil.

Financial invetments in new nuclear power plants make coal and natural gas powered electricity generation a better investment. The high cost of new nuclear electricity generation is due to the initial learning curve of a new technology. Government regulation of this new technology will eventually become more realistic, thus allowing nuclear electricity to become more competitive with coal and natural gas generated electricity. The United States must look at the financial and safety records of other countries that have commited their resources to nuclear electricity generation.

As fossil fuels become depleted, generation of electricity will become increasingly dependent upon nuclear energy, particularly for transportation and desalinization of water.

NUCLEAR ENERGY SAFETY

Nuclear energy, compared to other forms of energy must be considered to be in its infancy. Accident and incident reporting in the nuclear industry is well documented with three significant accidents in the 50-year history of civil nuclear power generation. These accidents have emphasized the dangers associated with nuclear power generation and have spearheaded the design of safer nuclear reactors.

THREE MILE ISLAND (USA 1979) where the reactor was severely damaged but radiation was contained and there were no adverse health or environmental consequences.

CHERNOBYL (Ukraine 1986) where the destruction of the reactor by steam explosion and fire killed 31 people and had significant health and environmental consequences.

FUKUSHIMA (Japan 2011) where three old reactors (together with a fourth) were written off and the effects of loss of cooling due to a huge tsunami were inadequately contained.

These are the only major accidents to have occurred in over 14,500 cumulative reactor-years of commercial operation in 32 countries.

The United States navy has over 160 ships that are powered by nuclear energy without a single major incident reported.

Nuclear energy research shows potential for nuclear reactors that are inherently safer than existing technology. These new generations of reactors are discussed in Section Eight.

NUCLEAR ENERGY ECONOMICS

At present, nuclear power is cost competitive with other forms of electricity generation, except where there is direct access to low-cost fossil fuels. There will come a point where oil depletion will cause the cost of oil to reach a level where oil derived products can no longer be used to power vehicles. At that time electric vehicles, home heating and industrial needs will increase the demand for electricity resulting in shortages of electricity and the need for nuclear electric power generation.

Fuel costs for nuclear plants are a minor proportion of total generating costs, though capital costs are greater than those for coal-fired plants and much greater than those for gas-fired plants.

Assessing the relative costs of new generating plants using different technologies is a complex matter and the results depend crucially on location. Coal is, and will probably remain, economically attractive in countries such as China, the USA and Australia with abundant and accessible domestic coal resources as long as carbon emissions are cost-free. Gas is also competitive for baseload power in many places, particularly using combined-cycle plants, though rising gas prices have removed much of the advantage.

Nuclear energy is, in many places, competitive with fossil fuels for electricity generation, despite relatively high capital costs and the need to account for all waste disposal and decommissioning costs. If the social, health and environmental costs of fossil fuels are also taken into account, the economics of nuclear power are outstanding.

NUCLEAR ENERGY WASTE DISPOSAL

Used Nuclear Fuel is currently stored at the nation's nuclear power plants in steel-lined, concrete pools or basins filled with water or in massive, airtight steel or concrete-and-steel canisters.

Recycling Used Nuclear Fuel will be practical when federal government develops advanced recycling technologies to take full advantage of the unused energy in the used fuel and reduce the amount and toxicity of byproducts requiring disposal.

Congress passed legislation in 1982 directing the U.S. Department of Energy (DOE) to build and operate a deep geologic repository for used nuclear fuel and other high-level radioactive waste at Yucca Mountain, Nev., a remote desert location. A year later, President Obama announced plans to discontinue the Yucca Mountain project and empanel a blue ribbon commission to provide recommendations for long-term management of high-level radioactive waste. DOE announced formation of the commission on Jan. 29, 2010.

NUCLEAR ENVIRONMENTAL CONSIDERATIONS

There are many environmental benefits from the use of nuclear energy. Nuclear power plants aid compliance with the Clean Air Act of 1970, which set standards to improve the nation's air quality.

Federal, state and local policymakers increasingly recognize nuclear energy's contribution to meeting growing electricity demand while reducing greenhouse-gas emissions. Nuclear power plants do not emit criteria pollutants or greenhouse gases when they generate electricity. The life-cycle emissions from nuclear energy are comparable to other non-emitting sources of electricity like wind, solar and hydropower.

Nuclear energy has one of the lowest impacts on the environment of any energy source because it does not emit air pollution, isolates its waste from the environment and requires a relatively small amount of land.

Clean electricity for transportation will reduce air emissions from the transportation sector by developing electric vehicles that can run farther and longer between charges. Clean electricity from nuclear plants can make these vehicles truly clean.

SECTION SIX
WORLD ENERGY STATISTICS

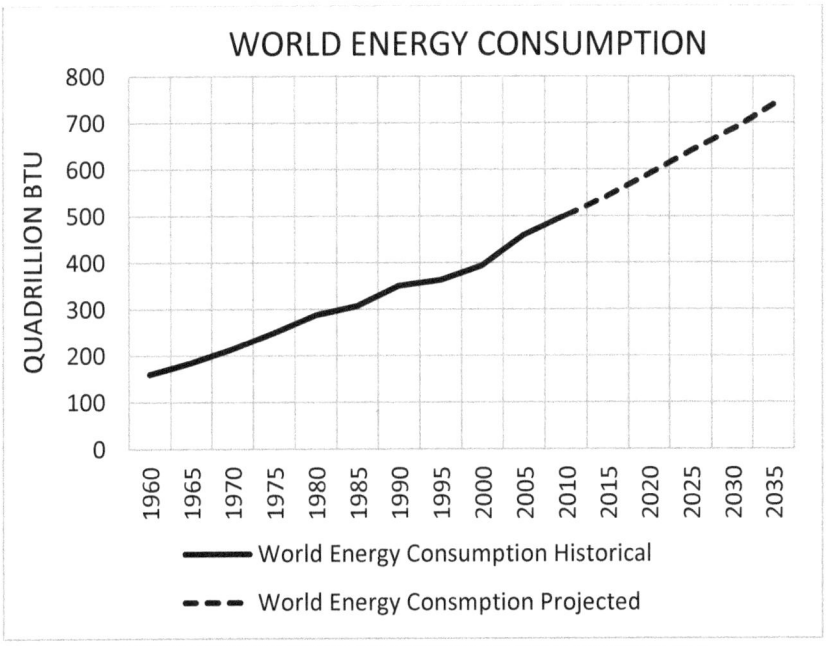

US ENERGY INFORMATION ADMINISTRATION HISTORICAL TABLE 11.1
US ENERGY INFORMATION ADMINISTRATION PROJECTED TABLE A1

World energy consumption increased from 215 quadrillion BTUs in 1970 to 510 quadrillion BTU in 2010, equivalent to 237% in 40 years. World population in the same 40 year period increased from 3.713 billion people to 6.852 billion people, equivalent to 184%. This computes to an increase of 28% in the average number of BTUs consumed per person for that same period. This increase can be attributed to advancements in technology which makes it possible for more people to afford more energy consuming devices such as automobiles, computers and electric tooth brushes. There are no statistics to indicate a change in this trend.

From 1975 to 1990 world total energy consumption increased at the rate of 1.91% per year while world total oil consumption increased at the rate of only .84% per year for the same period. From 1990 to 2005 world total energy consumption increased at the rate of 1.69% per year while world total oil consumption increased at the rate of only .18% per year for the same period. These diminished rates of increase in oil consumption can be attributed to concerns about depleting oil reserves which resulted in the growth of other forms of energy such as nuclear, wind, solar, and biofuels.

Although oil consumption is growing at a rate considerably less than total energy consumption, we do not know whether oil consumption can be further controlled in time to prevent an economic disaster when world oil reserves reach a critically low point.

The following table includes fossil fuels, nuclear energy, hydroelectric, geothermal and other renewable sources of energy. Other renewable energy is further broken down in a later table. To facilitate comparisons and trends, all forms of energy are shown in terms of BTU. Conversion tables are included in the appendix.

WORLD ENERGY CONSUMPTION OF	2000 Quadrillion BTU	2007 Quadrillion BTU	% increase in 7 years
Crude Oil	146.800	156.000	6
Coal	89.100	132.700	48
Natural Gas	90.900	109.000	19
Hydroelectric	26.700	29.600	10
Nuclear Energy	25.600	27.100	5
Natural Gas Liquids	9.600	11.900	25
Wind Power	.520	2.803	440
Bioethanol	.331	1.102	234
Geothermal	.239	.298	25
Other Renewable	Est .700	.989	41
World Total	390.490	471.490	21

Note: The previous table shows Natural Gas and Natural Gas Liquids as separate items. When gas comes out of the ground it contains several gases which are separated at the well site. Natural Gas is primarily Methane. Natural gas liquids (sometimes referred to as pipeline liquids) are the byproduct of drilling for natural gas. These gases include: ethane, propane, butane, and other gases. Further discussion of Natural Gas Liquids is included later in this section.

As of 2007, all energy consumed in the world was 471.5 quadrillion BTU. Energy from fossil fuels which includes Coal, Natural gas, Crude Oil and Natural Gas Liquids accounted for 86% of the world total. Renewable energy including Nuclear accounted for only 14% of the world total. For the period from 2000 to 2007 world energy consumption increased by 81 quadrillion BTU. For the same period, renewable energy including nuclear increased by only 7.8 quadrillion BTU which is less than 10% of the total increase in energy consumption. Based on these figures, we must conclude that our efforts to replace fossil fuels with renewable energy are not the answer to our depleting fossil fuel reserves.

Wind power and bioethanol show large increases of 440% and 234% respectively, which account for only 3.054 quadrillion BTU, which is less than 1% of total energy consumption. Considering that both wind power and bioethanol require large amounts of land, and that the best economically practical land for each has already being utilized, the trends of 440% and 234% respectively cannot continue indefinitely.

For the period 2000 to 2010, bioethanol consumption increased from .331 quadrillion BTU to 1.606 quadrillion BTU, a 385% increase.

For the period 2000 to 2010, wind energy consumption increased from .520 quadrillion BTU to 5.810 quadrillion BTU, a 1017% increase.

For the period 2000 to 2010, geothermal energy production increased from .239 quadrillion BTU to .328 quadrillion BTU, a 37% increase.

WORLD ENERGY CONSUMPTION
TRILLION BTU PER YEAR FROM OTHER RENEWABLE
SOURCES

OTHER WORLD ENERGY CONSUMPTION	2007 Trillion BTU
Solar Photovoltaic	41
Solar Thermal	3
Tide Wave Ocean	2
Municipal Waste	198
Industrial Waste	28
Solid Biomass	556
Biogas	107
Liquid Biofuels	11
Biodiesel	43
World Total	989

Sources of renewable energy listed above include some sources, such as Municipal Waste, Industrial Waste and Solid Biomass that have been around for many years. Total of all of the above sources of renewable energy (989 trillion BTU) account for only 0.2% of total world energy consumption.

Solar Photovoltaic and Solar Thermal combined accounted for only .01% of world energy consumption. Therefore we must conclude that contrary to many predictions, renewable solar energy is not the long term answer to consumption of our fossil reserves. When President Obama made loan guarantees of $535 million to Solyndra, the solar panel manufacturer, he did not understand the small part that solar energy contributes to our enormous energy consumption, nor did he understand that mobility fuels are our immediate problem.

Energy consumption will continue to increase as long as population increases and technology provides us with new ways to consume more energy per capita, or, until there are no longer sufficient sources of energy to support the extravagant life styles that the developed countries have become accustomed to. Although attempts are being made to control population throughout the world, particularly in China, population continues to increase at an alarming rate. Graphs in the section on Population show that attempts to control population are not working.

Feeble attempts are being made by world governments to reduce energy consumption of individuals by legislating better gasoline mileage for automobiles and encouraging more efficient light bulbs. These are merely a drop in the bucket. The graph showing World Energy Consumption per Capital indicates that attempts to reduce energy consumption is not working.

Can the increases in energy consumption shown in the chart at the beginning of this section continue at these rates indefinitely? NO. Will alternative energy be sufficient and timely to prevent depletion of our fossil fuels? PROBABLY NOT

Energy is usually expressed in terms of BTU. The British Thermal Unit (BTU) is the traditional unit of energy used around the world. It is the amount of energy needed to heat 1 pound of water 1 degree Fahrenheit. A BTU is an accumulation of energy used over a period of time, usually shown for a period of one year.

Examples of energy consumption in terms of BTU are a 100 watt light bulb burning for 10 hours which will consume 3412 BTU, or one gallon of gasoline which contains 120,000 BTU.

For the year 2008 the world consumed energy from all sources equivalent to 490,820 Trillion BTU.

Prior to the 20th century there was never any thought that technology would provide the population with so many means of consuming world natural resources. Very few people understand that technology is taking us in a direction that eventually will cause the depletion of our natural resources resulting in starvation and suffering for most of mankind.

Americans: Wake up and visualize what the world will be like in 50, 100 or 200 years if we pursue our present path of energy extravagance.

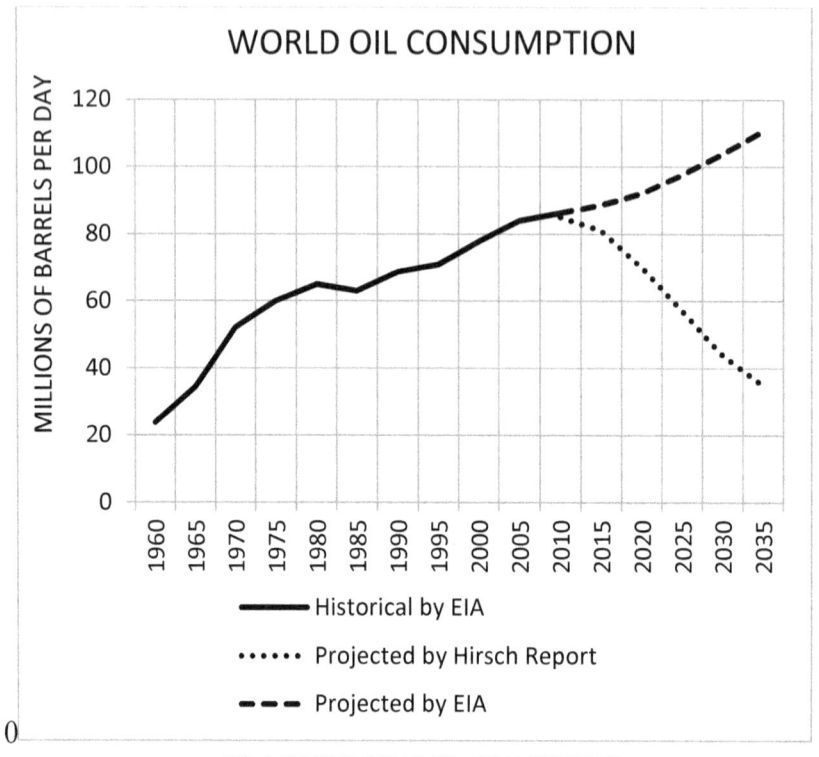

WORLD OIL CONSUMPTION

U.S. ENERGY INFORMATION ADMINISTRATION
INTERNATIONAL ENERGY STATISTICS DATABASE TABLE A5

The previous graph (dash line) shows world oil consumption in 2010 at 88.7 million barrels per day, increasing to 110.6 million barrels per day by 2035. This prediction from the U.S. Energy Information Administration is based on a linear extrapolation of world oil consumption from 1970 to 2010 without taking into consideration oil depletion which will cause prices to increase resulting in lower oil consumption.

Contrary to the U.S. Energy Information Administration prediction is the Hirsch Report (dotted line) commonly referred to the report "Peaking of World Oil Production: Impacts, Mitigation, and Risk Management" which shows oil production peaking around 2010 and then decreasing through 2035. Contributors to the Hirsch Report include many scientists, geologists, economists, universities and the World Energy council.

Further credibility is given to the Hirsch Report when we examine how our Alaska oil production peaked. Americans believed that those "enormous" amounts of oil discovered in Alaska would last forever, so we built an expensive pipeline to bring oil to the port of Valdez where it could be shipped to the United States to fuel our gas guzzlers. Alaska oil peaked in 1988 at 2,000 thousand barrels per day and has declined 65% since reaching peaking in 1988. By 2008, Alaska oil was down to only 700 thousand barrels per day.

In 2009 the World consumed 1.269 billion gallons of oil (82.8 million barrels per day). Below are some of the largest consumers of that oil.

2009 WORLD OIL CONSUMPTION	Million Barrels per Day	% of World Total
World Oil Consumption	82.769	100.0
U.S. Oil Consumption	19.203	23.2
U.S. Transportation Total	13.556	16.4
U.S. Cars & Light Trucks	8.712	10.5
U.S. Heavy Trucks	2.900	3.5
U.S. Industrial	4.705	5.7
U.S. Residential	3.174	3.8
U.S. Commercial	2.113	2.5
World Ships	7.723	9.3
China	8.200	9.9
Japan	4.363	5.3
India	2.980	3.6
Russia	2.740	3.3
Brazil	2.460	3.0
Germany	2.437	2.9
Saudi Arabia	2.430	2.9

2009 WORLD OIL PRODUCERS	Million Barrels per Day
Saudi Arabia	10.584
Russia	9.285
United States	7.241
Iran	4,081
Mexico	3.824
China	3.490
Norway	3.188
Canada	3.085
Venezuela	2.980
United Arab Emirates	2.667
Nigeria	2.508
Kuwait	2.424
United Kingdom	2.029
World Total	82.769

2008 WORLD OIL RESERVES	Billion Barrels Crude Oil
Venezuela	297
Venezuela Oil Sands	513
Saudi Arabia	267
Canada Conventional	178
Canada Oil Sands	173
Iran	136
Iraq	143
Kuwait	104
United Arab	98
Russia	74
Libya	44
Nigeria	36
Kazakhstan	30
Qatar	25
United States Conventional	21
United States Oil Sands	32
World Total	2,103

DATA AS OF JANUARY 1, 2009
US ENERGY INFORMATION ADMINISTRATION HISTORICAL TABLE 11.4

WORLD OIL PRODUCTION COSTS

There are vast amounts of oil in the world and there are vast demands for production of that oil. When the world entered the industrial revolution oil was easy to produce. Oil producers looked for oil on the surface and simply drilled down a few feet until oil gushed out of the ground. Oil was cheap to produce until those wells became depleted. Replacement wells had to be drilled deeper and in less accessible areas. The cost of oil from the new wells was more expensive, but with all the new uses for oil it was a bargain.

Cost of producing oil in the United States became more expensive than importing oil from countries such as Saudi Arabia until world oil prices increased to the point that made it economically practical for the United States to produce more oil from less accessible locations and shale reserves.

VENEZUELA OIL RESERVES AND PRODUCTION

Proven oil reserves in Venezuela are claimed to be the largest in the world, according to an announcement in early 2011 by President Hugo Chavez and the Venezuelan government reporting proven reserves of 297 billion barrels surpassing that previously reported by Saudi Arabia in 2009 claiming proven reserves of 265 billion barrels.

Venezuela is one of the largest suppliers of oil to the United States. In 1996 United States imported 1.876 million barrels per day to the U.S. In 2010 United States imports from Venezuela were down to .987 million barrels per day.

Venezuela is also a major oil refiner and the owner of the Citgo gasoline chain.

United States trade relations with Venezuela as of 2012 were good, but diplomatic relations were poor. This is another reason why the United States must become energy independent and not depend upon this type of relationship for our critical oil.

SAUDI ARABIA OIL RESERVES AND PRODUCTION

Saudi Arabia produced 10.3 million barrels per day in 1980, 10.6 million barrels per day in 2006, and approximately 9.2 million barrels a day in 2008. Saudi Arabia maintains the world's largest crude oil production capacity, estimated to be approximately 11 million barrels per day in 2008 and announced plans to increase this capacity to 12.5 million barrels per day by 2009. Cumulative production up to, and including, the year 2009 was 119.4 billion barrels. Based on 296 billion barrels of reserves, past production amounts to 40% of reserves.

Based on recent oil sands discoveries in Canada, Venezuela and the United States, Saudi Arabia no longer has the greatest oil reserves, but it does maintain the greatest oil production capacity because of the ease of producing Middle East oil.

CANADIAN OIL RESERVES, PRODUCTION

Canada is the sixth largest oil producing country in the world. In 2008 it produced an average of 2.750 million bbl per day. Most of Canadian petroleum production was exported, almost all of it to the United States. Canada is the largest single source of oil imports into the United States.

Oil reserves in Canada were estimated at 179 billion barrels in 2007. This figure includes oil sands reserves which are estimated by government regulators to be economically producible at current prices using current technology. According to this figure, Canada's reserves are third only to Venezuela and Saudi Arabia. Over 95% of these reserves are in the oil sands deposits in the province of Alberta. Over 99% of Canadian oil exports are sent to the United States, and Canada is the United States' largest supplier of oil.

The Athabasca Oil Sands in the Provence of Alberta is the largest producer of conventional crude oil, synthetic crude, natural gas and gas products in Canada. The Athabasca Oil Sands have estimated oil reserves in excess of that of the rest of the world, estimated to be 1.6 trillion barrels. With current technology, 315 billion barrels are recoverable.

Hibernia is an oil field in the North Atlantic Ocean, 196 miles east-southeast of St. John's, Newfoundland, Canada. Proven and probable estimated reserves are 1.395 billion barrels of oil. Production is from a single platform *"Hibernia"*, the world's largest oil platform. Production commenced November 1997, with initial production rates in excess of 50,000 barrels of oil per day. Total production as of August 2010 was 704 million barrels

Canadian Oil Production in million barrels per day average in 2008

Athabasca Oil Sands Production	1.300
Hibernia Production est.	.200
Other	1.250
Total Production	2.750
Total Export	1.780

Canadian total oil production is estimated to increase to over 4 million barrels per day by 2020. Interests in Canadian oil production are held by China, USA, UK, Netherlands, France, Japan, Norway and Korea. Eventual consumption of Canadian oil will depend upon the Keystone XL pipeline, proposed pipelines to west coast shipping ports, fuel emission standards, and politics.

UNITED STATES OIL PRODUCTION AND CONSUMPTION

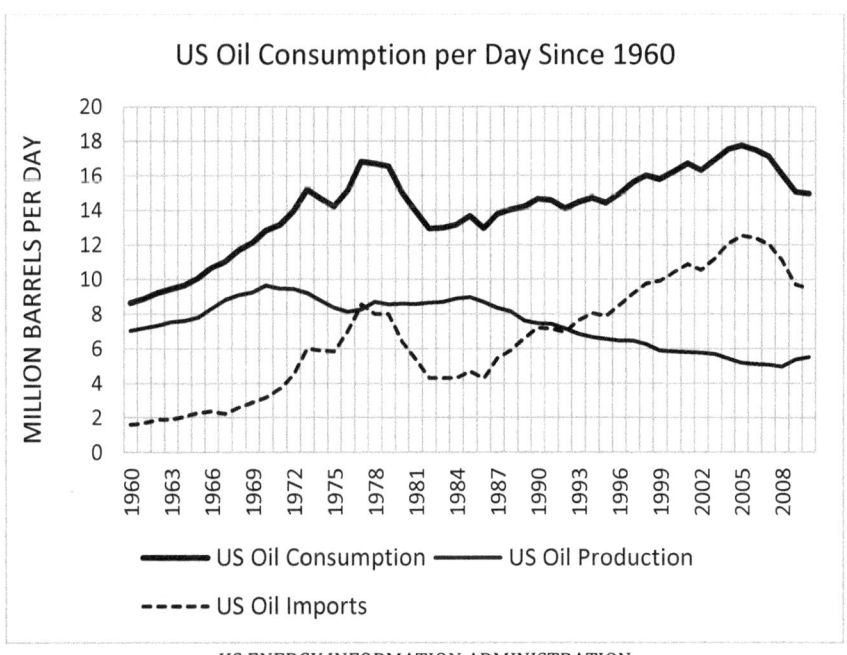

US ENERGY INFORMATION ADMINISTRATION
INDEPENDENT STATISTICS & ANALYSIS TABLE 5.1b

United States *oil imports* in million barrels/day (average) in 2010. Note that the United States does not import any oil from Russia although their production is second in the world almost on par with that of Saudi Arabia. This is due to both political and technical reasons which could change in the next decade.

Canada	2.532
Mexico	1.280
Saudi Arabia	1.084
Nigeria	1.025
Venezuela	.987
Other	4.845
Total Imports	11.753

United States *petroleum products exported* in million barrels/day (average) in 2010. Export of unprocessed crude oil in 2010 is estimated at only 42,000 barrels per day. The following figures are for petroleum products such as gasoline and diesel which the United States exported to countries that do not have adequate if any refinery capacity.

Mexico	.447
Canada	.211
Netherlands	.160
Brazil	.122
Unprocessed Petroleum	.042
Other	1.330
Total Exports	2.312

United States *oil net imports* in million bbl/day (average) in 2010

Total Imports	11,753
Total Exports	2,312
Net Imports	9.441

United States *oil production* in million bbl/day (average) in 2010

Onshore Crude Oil	3.691
Offshore Crude Oil	1.812
Natural Gas Liquids	2.074
Total Production	7.577

United States *oil consumption* in million bbl/day (average) in 2010. These figures do not include crude oil that is refined into petroleum products and then exported.

> Total Consumption   17.018
> Total Production   7.577
> Net Imports   9.441

United States *oil production* has been decreasing since 1970 when it peaked at over 9.637 million barrels per day. This trend may temporarily reverse if the Bakken Formation in Montana and North Dakota meets estimates of 1.0 million barrels per day by 2035. The following tabulated estimates are based on Bakken increasing production from 2010 to 2035 by .542 million barrels per day and Eagle Ford Shale Formation in Southeast Texas increasing production to .420 barrels per day

> Total Consumption est. 2035   20.250
> Total Production est. 2035   8.539
> Net Imports 2035   11.711

Based on these figures, newly developed shale oil reserves will not reduce United States dependence on foreign oil imports unless consumption is reduced and renewable energy such as Algae fuel is produced in sufficient quantities to meet increasing demands.

United States production of oil comes from numerous oil fields scattered throughout the lower 48 states, Alaska and offshore wells. Some of the largest concentrations of oil production are described in the following paragraphs.

East Texas Oil field discovered in 1930 was the largest oil field of its time. It originally contained 7 billion barrels of oil and as of 2010 has produced 5.2 billion barrels of oil.

Spraberry Trend in West Texas started drilling operations in 1943. In 2007, the U.S. Department of Energy ranked The Spraberry Trend third in the United States by total proved reserves, and seventh in total production. Estimated reserves exceed 10 billion barrels and by the end of 1994 the field had reported a total production of 924 million barrels. The Spraberry Trend retains about 90% of its original calculated reserves, largely due to the difficulty of recovery. Shale rocks make up a total of 87% of the Spraberry. Because most of the initial wells showed precipitous and mysterious production declines, the area was dubbed "the world's largest unrecoverable oil reserve. As of 2009, there were approximately 9,000 active wells in the Spraberry Trend.

Prudhoe Bay Oil Field on Alaska's North Slope was the largest oil field in the United States, originally estimated to contain 25 billion barrels as of 1977. As of 2006, remaining recoverable oil was estimated at 2 billion barrels. Production at peak in 1979 was 1.5 million barrels a day. As of 2010 production was down to only 660,000 barrels per day.

Kuparuk Oil Field located in North Slope Borough, Alaska was discovered 1969. Production began 1981 and production peaked in 1992 at 322,000 barrels per day The Kuparuk Oil Field, now produces approximately 230,000 barrels per day of oil and is estimated to have 2 billion barrels of recoverable oil reserves.

Offshore oil from all United States OCS (Outer Continental Shelf) during 2002 was 1.650 million barrels per day. This decreased to 1.615 million barrels per day in 2010 of which Gulf of Mexico accounted for 1.550 million barrels per day.

The Bakken and the Eagle Ford are each expected to ultimately produce 4 billion barrels of oil. That would make them the fifth- and sixth-biggest oil fields ever discovered in the United States. The top four are Prudhoe Bay in Alaska, Spraberry Trend in West Texas, the East Texas Oilfield and the Kuparuk Field in Alaska. North Dakota's Bakken shale produced .006 million barrels per day in 2006, .020 million barrels per day in 2007, and .458 million barrels per day in 2010. Some analysts think it could produce a million barrels a day by 2035 and may contain billions of untapped barrels. However, production depends on economic considerations and controversial fracking technology.

Eagle Ford Shale in eastern Texas was one of the most actively drilled targets for oil and gas in the United States in 2010. The oil reserves are estimated at 3 billion barrels with potential output of .420 million barrels per day.

WORLD NATURAL GAS RESERVES AND CONSUMPTION

	Reserves Trillion Cubic Ft. Natural Gas	Production Trillion Cubic Ft. Per Year	Reserve Life Years
Russia	1,680	20.6	82
Iran	991	7.1	139
Qatar	891	2.7	330
Saudi Arabia	258	2.7	95
United States	2,600	20.9	123
United Arab	214	1.8	119
Venezuela	170	.8	212
World	6,261	110.4	56.7

CENTRAL INTELLIGENCE AGENCY-THE WORLD FACTBOOK - JANUARY 1, 2011

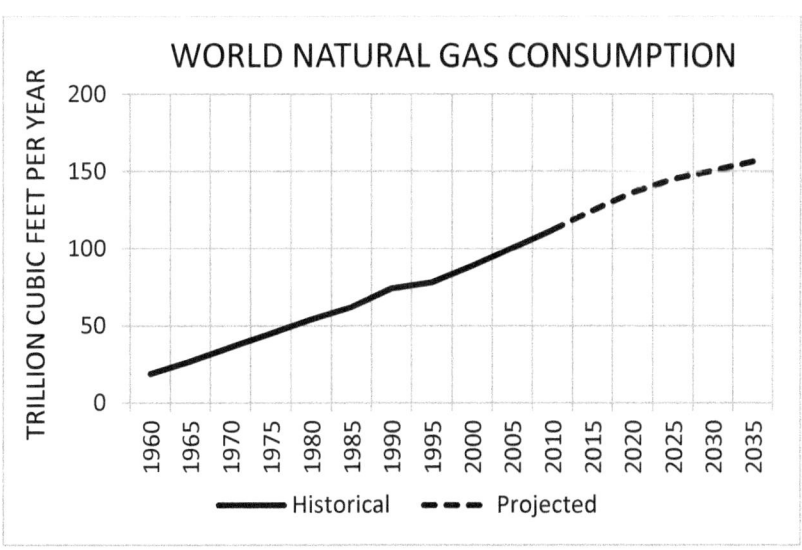

US ENERGY INFORMATION AGENCY JANUARY 2009
INTERNATIONAL ENERGY OUTLOOK 2010 TABLE A6

Estimates of United States natural gas reserves are very controversial. Section Eight, Future Energy has a detailed paragraph titled "US Natural Gas Controversy".

WORLD COAL RESERVES AND CONSUMPTION

WORLD COAL RESERVES	Billion Short Tones Coal	Production Million Tones Per Year	Reserve Life Years
United States	261	973	268
Russia	173	298	580
China	126	3050	41
Australia	84	409	205
India	65	557	116
Ukraine	37	74	500
South Africa	33	250	132
World	909	6,940	130

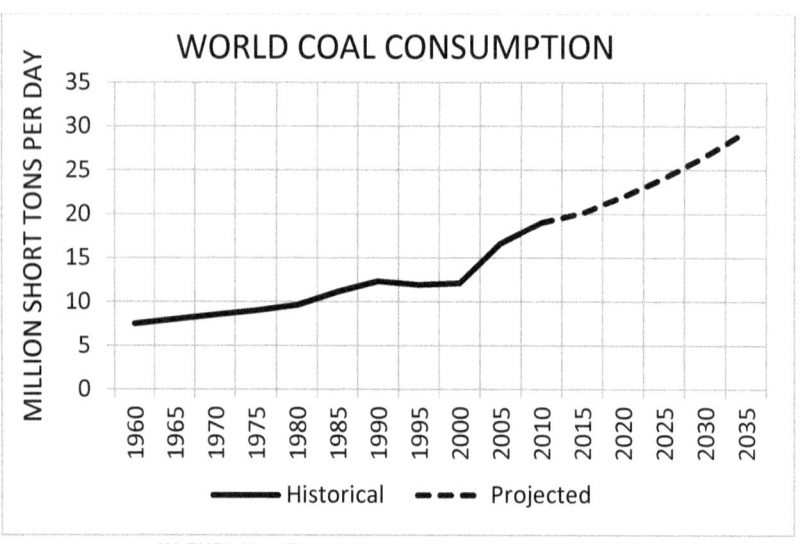

US ENERGY INFORMATION AGENCY JANUARY 1, 2009

Coal is the world's most abundant source of fossil fuel energy and produces the greatest amount of greenhouse gases. America's coal is used primarily for the production of electricity. Approximately one third of all CO_2 emissions due to human activity derive from fossil fuels used to generate electricity, with each power plant capable of emitting several tons of CO_2 annually. These emissions could be reduced substantially by capturing and storing the CO_2. If capture is used to minimize CO_2 emissions from a power plant it would add considerably to the cost of generating electricity. Once CO_2 is captured it needs to be stored securely for hundreds or even thousands of years. Major potential reservoirs have been identified and development is underway.

Most climate change recommendations have to do with the pollution caused by burning and processing of coal, and measures to limit the yearly output of greenhouse gases, for example placing a moratorium on new coal-fired power plants that do not sequester their carbon dioxide emissions.

As of 2010, natural gas is being used more frequently to power new electric generating plants because it is cleaner burning than coal. However, coal may continue to be the energy source for power plants if the controversy over extraction of natural gas by hydraulic fracturing is proven to contaminate our drinking water supply. Future selection of coal or natural gas to power new electric power generating plants will remain controversial until the problems associated with hydraulic fracturing are resolved.

New uses of coal include conversion of coal to gasoline by methods that are being economically practically. The Adams fork Energy Project is discussed in Section Eight, Future Energy.

WORLD AND US ELECTRIC CONSUMPTION

Electricity can be used to power everything that requires power. Because of the technical limitations of storage, it is not practical to use it for many transportation applications. However, with the advent of new compact high efficient batteries, electricity is becoming more practical for transportation. This is shown in the dramatic increase in hybrid vehicles and pure electric vehicles. With limitations, airplanes are flying with electric motors powered by modern batteries.

New uses of electricity require more electricity generation. It is too early to determine how new electricity to fill new requirements will be produced. New electricity could come from conventional coal or natural gas fired generators, wind power, solar energy, geothermal or nuclear. The answer to this is in the hands of our scientists and engineers, not our politicians.

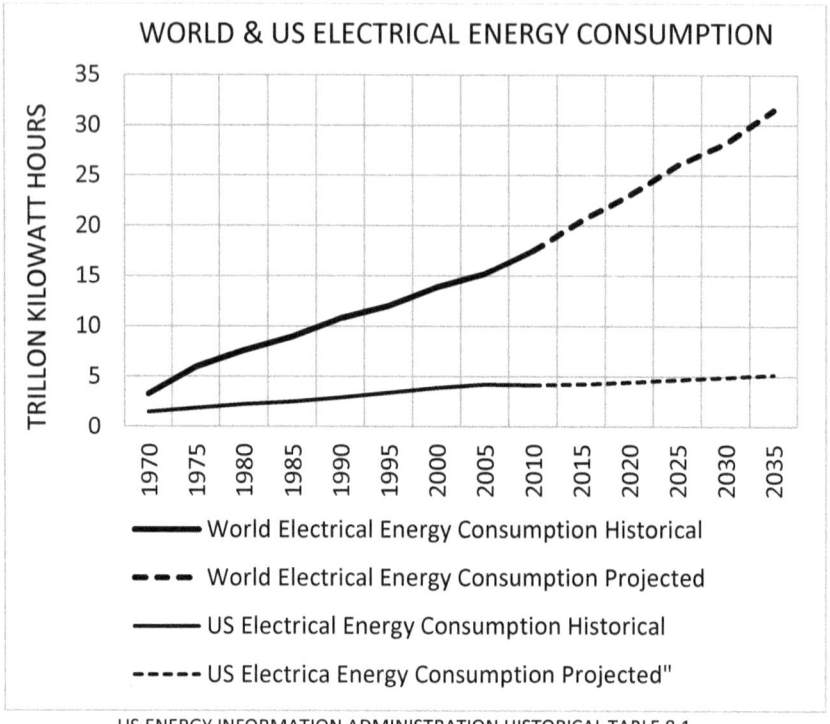

US ENERGY INFORMATION ADMINISTRATION HISTORICAL TABLE 8.1

WORLD AND US ELECTRICITY CONSUMER RATES
US CENTS/1 KWH

Argentina	5.70
Brazil	34.18
Canada	10.78
Chile	23.11
Denmark	40.38
France	19.40
Germany	34.50
Iceland	3.93
Italy	28.40
Israel	12.35
Philippines	30.45
United States average	11.20
United States Hawaii	25.20
United States Idaho	6.54

The preceding prices are for residential customers. Price charged to the consumer is made up of the cost of generation (approximately 60%), distribution (approximately 30%), and transmission (approximately 10%).

Large variations in prices are due to the method of generation. Countries with large amounts of hydroelectric or geothermal generators will have lower prices. Residential consumer rates will be constant throughout the day while industrial consumer rates will vary depending on the time of day. During off peak hours, large industrial consumers will pay minimum rates, but will pay larger rates during peak hours.

As electric vehicles become more popular, special rate structures will encourage charging of vehicles during off peak hours.

WORLD HYDROELECTRICITY

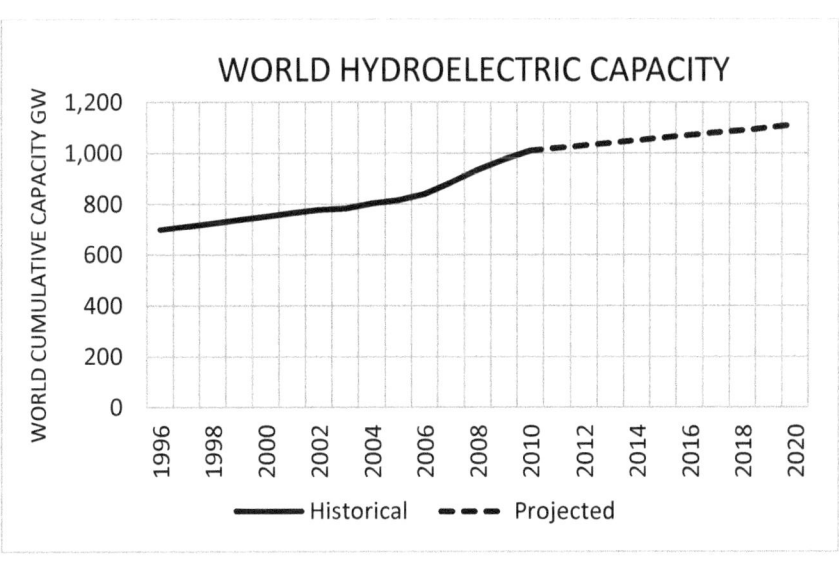

The above graph shows a large increase in hydroelectric capacity from 2003 through 2011 which is attributed to the two largest dams, Three Gorges Dam, China and the Itapúa Dam, between Brazil and Paraguay.

World hydroelectricity installed capacity as of 2009 was 1,010,000 megawatts. As with any resource development program, the most economically practical sites are selected first. Below is a tabulation of existing hydroelectric power throughout the world.

World Hydroe-lectricity 2009	2009 Installed Capacity MWe	2009 Actual Production Trillion BTU per year
China	197,000	2,225
Canada	89,000	1,259
Brazil	69,000	1,242
United States	79,500	853
Russia	45,000	570
Norway	27,500	478
India	33,600	392
Venezuela	14,600	293
Japan	27,200	235
Sweden	16,200	222
Other	411,400	
World Total	1,010,000	

BP STATISTICAL REVIEW 2009

The above table shows production adjusted for capacity factors which range from .37 for China to .67 for Venezuela. Installed Capacity or Rated Capacity is for short periods of time during periods of peak demand. When Installed Capacity is not required typically during night time operation, the output is referred to as Actual Output which could be as low as 30% of Installed Capacity.

World Largest Hydroelectric Projects	Installed Capacity MWe	Quad BTU per year
Hoover Dam US	2,080	.08
Grand Coulie US	6,809	.20
Aswan Dam Egypt	2,100	.06
Itaipu S.A.	14,000	.42
Three Gouges China	22,500	.67

The tabulation below lists future major hydroelectric projects throughout the world. Note that the United States has no major hydroelectric projects planned due to the lack of economically practical sites.

Note that future planned hydroelectric projects are only 10.3% of existing power measured by installed capacity.

	Future Projects	Installed Capacity MWe
China	Xiluodu Dam	12,600
	Xiangjaba Dam	6,400
	Nuozhadu Dam	5,850
	Jimping 2	4,800
	Jimping 1	3,600
	Pubugou Dam	3,300
	Gouptan Dam	3,000
	Guanyinyan Dam	3,000
	Lianghekou Dam	3,000
	Dagabgsgan Dam	2,600
	Guandi Dam	2,400
	Liyuan Dam	2,400
	Ludila Dam	2,100
	Shuangjangkou Dam	2,000
	Ahai Dam	2,000
	China Total	59,050
Brazil	Belo Monte Dam	11,180
	Jirau Dam	3,300
	Santo Antonio Dam	3,150
	Brazil Total	17,630

India	Siang Upper HE	11,000
	Lower Subansin Dam	2,000
	India Total	13,000
Burma	TaSang Dam	7,110
Russia	Boguchan Dam	3,000
Vietnam	So'n La Dam	2,400
Venezuela	Tocoma Dam	2,160
	World Total	104,350

Before a dam, hydroelectric generators and distribution lines can be built, the project must be proven economically practical. An example is the Congo River and the Inga Falls in the middle of Africa capable of producing 40,000 megawatts of electricity, twice that of the Three Gorges Dam, the worlds largest hydroelectric project. The dam would be built in an area that has the lowest electric energy per capita in the world. However, due to political and ecoonomic problems, construction is not yet planned.

RATINGS OF POWER PLANTS including hydroelectric, coal, natural gas, geothermal and nuclear which connect to the electric grid system have the capability of producing Installed Capacity in MWe (megawatts of electtricity), also referred to as Nameplate Capacity or Rated Capacity for short periods of time during periods of peak demand. When Installed Capacity is not required typically during night time operation, the output is referred to as Actual Output which could be as low as 30% to 50% of Installed Capacity. Actual output will vary depending upon demand from the Electric Grid which will select power sources based on economic considerations. Further detailed discussion of the Electric Grid in Section Seven.

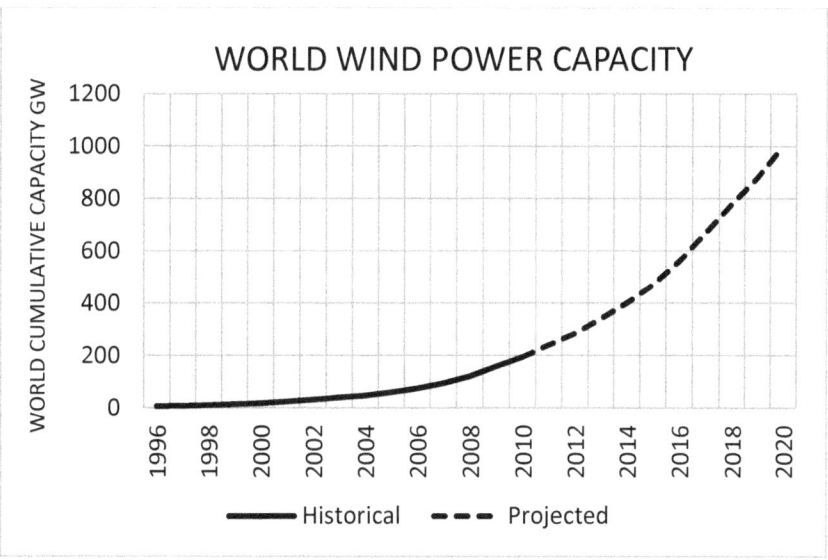

GLOBAL WIND ENERGY COUNCIL

Wind power is the fastest growing source of energy in terms of actual capacity with a growth rate over the last 10 years of almost 30% per year. Electrical energy produced by wind power is competitive with electricity produced by fossil fuels at 2009 fuel costs. As fuel costs increase, wind power will become increasingly more competitive. Technology advances and production learning curves are contributing to make wind power more attractive each year.

An advantage of wind power compared to solar energy is the higher capacity factor of wind power which allows connection to the electrical grid for long periods of time without the need for storage.

The most important factor when selecting locations for wind power is consistent favorable winds. Areas in the United States such as offshore northeast, offshore California and the west central states provide the necessary consistent favorable winds. However, some areas such as west central United States where wind is favorable will require additional long distance high voltage transmission lines.

Some offshore locations for wind turbines cause adverse reaction from residents with homes facing the water who take the position of "not in my back yard".

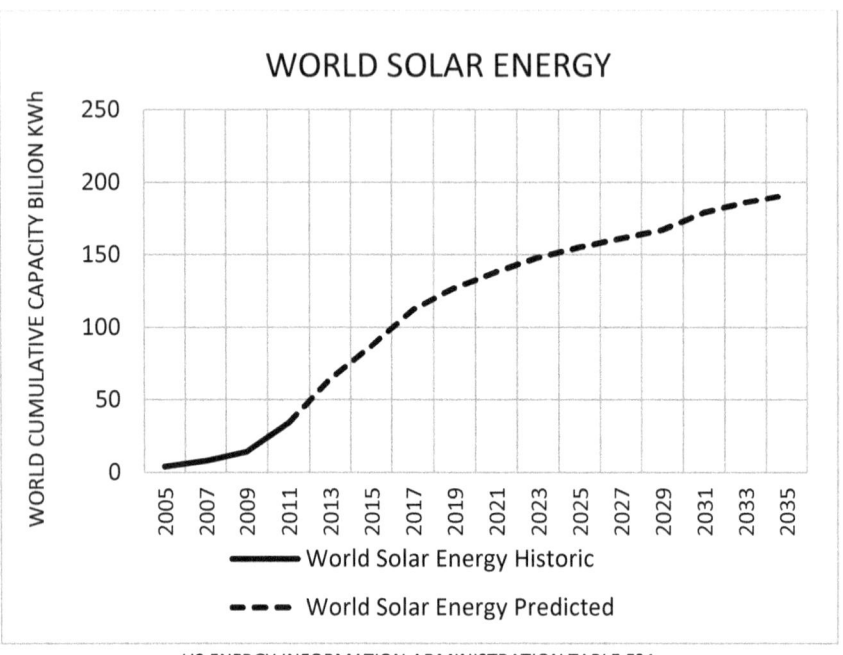

US ENERGY INFORMATION ADMINISTRATION TABLE F21

Solar energy shown in the above table is based on central pro-duction facilities, not including individual solar panels on private homes and businesses which is difficult to estimate and would increase these estimates by less than 5%.

A disadvantage of solar energy is the need to store energy while the sun is in a position to produce the energy, and convert it into electricity when it is needed at a later time.

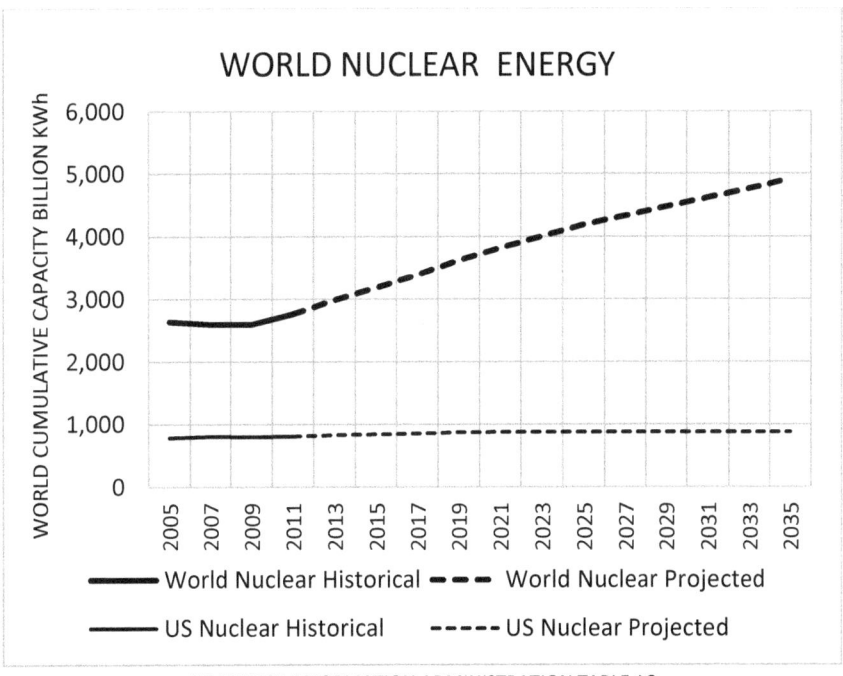

US ENERGY INFORMATION ADMINISTRATION TABLE A8

	Operating	Under Construction	Planned
Canada	18	2	4
China	13	27	50
France	58	1	1
Germany	17	0	0
India	20	4	20
Japan	55	2	12
South Korea	21	7	4
Russia	32	10	14
United Kingdom	19	0	4
United States	104	3	9
World Total	435	63	150

The above figures are prior to the Fukushima disaster and do not list countries with less than 12 reactors.

SECTION SEVEN
ENERGY STORAGE AND DELIVERY

NECESSITY TO STORE ENERGY

When we flip on a light switch, we expect the lights to come on, and they do. Electricity in the United States is extremely dependable and economical when compared to other nations around the world. To accomplish this, the time factor of when electricity is produced and when it is consumed make it economically desirable to store energy so that it is available when needed. There are no precise figures available that show when there is the greatest and least demand for energy. During the day when factories are operating and home appliances are running, there is a greater demand for electric energy compared to the middle of the night when most people are sleeping.

Lights come on because there are electrical generators powered by coal, natural gas, oil, hydroelectric, wind, geothermal or nuclear. The vast electrical generation and distribution system is controlled by many computers on a network of power lines, transformers and substations. Under extreme conditions, electricity may be consumed a thousand miles from where it is generated

However, there may not be enough power from all the electrical generators in the system under conditions that could occur periodically such as at peak demand hours. When peak demand for electricity occurs, the network will call on pumped hydraulic storage facilities and in the extreme case will call on less efficient gas turbine generators.

The time factor of when and where energy, particularly electricity is availability and when it is consumed make it economically desirable to store energy so that it is available when needed. During the day when factories are operating and home appliances are running, there is a greater demand for electric energy compared to the middle of the night when many factories are shut down and most people are sleeping.

KINETIC AND POTENTIAL ENERGY

In brief, kinetic energy by definition is momentum while potential energy is stored energy.

Electricity is a form of kinetic energy because there are electrons in motion. An automobile in motion is a form of kinetic energy because there is mass in motion. Heat is kinetic energy because there are molecules in motion. Kinetic energy in these examples all have in common, the difficulty, if not the impossibility of storage.

Potential energy exists in the Earth's fossil reserves, trees in the forest, water at a high elevation, compressed air, a roller coaster at the top of the track and many more. All forms of potential energy can be considered as energy in storage waiting to be converted into kinetic energy.

Kinetic energy in the form of solar energy can be used to generate electricity using photo electric cells, or by heating water which can be can be stored and later used to drive turbines connected to generators that produce electricity.

Kinetic energy is the power in any vehicle or machine such as a roller coaster, train, truck, car, boat, or simple flywheel. The earth has vast amounts of kinetic heat energy below the surface referred to as geothermal energy. Scientists believe that geothermal energy has potential to provide clean, safe electrical energy that may in the future solve much of our energy problems.

CREATION OF ENERGY

The universe was created with a finite amount of energy. Man cannot create nor destroy that energy, but he can convert it from potential energy to kinetic energy and back and forth.

It may appear that we are creating energy when we grow ethanol fuel. Answer is that we are taking sun energy and converting it into liquid energy. Growing ethanol fuel could be the answer to world energy needs if not for the practical aspects of the amount of land, water and electric energy required to grow ethanol fuel. Growing ethanol fuel will continue to be controversial as a compromise between fuel for transportation and fuel (food) for human nutrition.

WHEN AND WHERE NEEDED

Public utilities are obligated to supply their consumers electrical energy when it is required and where it is required, not when or where it is convenient for the utilities to produce the electricity. This is the reason that we have many high voltage electric transmission lines crossing the United States apparently in a random fashion. The subject of producing, selling, buying and consuming electricity is controversial and the reason for the scandals and eventual collapse of the Enron Corporation. In order to understand why electricity is transmitted long distances, it is necessary to understand consumer needs for baseload power and peak power.

BASELOAD POWER

Baseload is the minimum amount of power that a utility or distribution company must make available to its customers, or the amount of power required to meet minimum demands based on reasonable expectations of customer requirements. Baseload values typically vary from hour to hour in most commercial and industrial areas. A typical household refrigerator operating 24/7 is a good example of a constant baseload demand for electricity.

PEAK POWER

In the United States, peak power demand often occurs in the afternoon, especially during the summer months when the air conditioning load is high and people return home from work, start cooking dinner, and turn up the air conditioning. During this time many workplaces are still open and consuming power.

Peak power requirements must be met by the electric utilities so as not to cause a brown out or possibly a black out. In order to meet peak demands when and where they are required, peaker plants are placed strategically on the electrical grid. A peaker plant may operate many hours, a day, or as little as a few hours per year, depending on the condition of the region's electrical grid. It is expensive to build an efficient power plant, so if a peaker plant is only going to be run for a short or highly variable time, it does not make economic sense to make it as efficient as a base load power plant. Peaker plants that burn natural gas in turbines are generally less efficient than large coal or natural gas power generators, but, the low initial cost of construction of a peaker plants make them economically practical.

CAPACITY FACTOR

Machines that generate electricity as well as machines that consume electricity have a rating in terms of horsepower or wattage. The rating is in terms of the maximum sustainable power that the machine can safely produce or consume. These ratings are referred to as nameplate capacity. As an example, a wind turbine may have a nameplate rating of 1 megawatt, but will seldom produce that amount of power due to variations in wind velocity. Actual output will be the total power output for a finite length of time, usually one year. If that particular wind turbine produced a total of 2,000 MWh (megawatt hours) of electricity but had the capability to produce 8,760 MWh if wind conditions were perfect all year round, that wind turbine would have operated at a capacity factor of 22.8 % for the year.

Estimates have been made for electricity production in the entire world providing the following capacity factors:

Wind Power	20-40%
Hydroelectric	30-80%
Nuclear World	60-100%
Nuclear US	92%
Photovoltaic – Arizona	19%

Compare the capacity factors between a photovoltaic farm in Arizona and a wind turbine farm in Texas. The photovoltaic farm may produce close to 100% of nameplate capacity for a short period of time every day while the sun is directly overhead and is not obstructed by clouds, giving it a relatively low capacity factor of only 19%. Maximum power output occurs at noon when demand for electricity is not at a peak, thus necessitating other electricity generation such as coal fired power plants to fill in at peak demand times. The wind turbine farm in Texas may produce 100% of nameplate capacity only once a year, probably at a time when demand is not at a peak.

The cost of construction and capacity factor of wind turbine farms compared to photovoltaic farms is the reason that wind power has been growing at a rate of 30% a year. It is predicted that if present trends continue, wind energy will surpass nuclear generation and hydroelectric generation, individually by the year 2020.

ELECTRIC POWER TRANSMISSION

As we deplete our fossil fuels, electricity generated from wind turbines, nuclear and geothermal power plants will become our primary source of electrical energy. These electricity generators may be located long distances from where the electricity will be consumed. For safety reasons future nuclear power stations will be located in remote areas so that if radiation is released it will have the least effect on populated areas. Although geothermal energy is available less than 5 miles below the surface in most areas, economic factors will determine where new geothermal electricity generation facilities will be located.

Electricity is available from coal fired generators, natural gas fired generators, hydroelectric dams, pumped hydroelectric power plants, geothermal electric generators, wind generators, solar generators, turbine peak power generators, and nuclear power plants. Some of these generators are close enough to the consumer so that the electricity can be distributed on medium voltage lines of less than 50 kV, then to pole mounted transformers reducing the voltage to less than 1 kV, and again to low voltage used in households.

When electricity must be transmitted greater than 50 miles, voltages greater than 100 kV are used to reduce energy loss. These overhead transmission lines have large towers that can be seen for many miles. Longer distance transmission up to 2,500 miles is typically done with overhead lines at voltages from 100 kV to 1,200 kV. Still greater distances up to 4,300 miles utilize DC with voltages up to 800 kV.

Transmission and distribution losses in the USA were estimated at 6.5% in 2007.

Generation, transmission, distribution and control centers make up the United States electricity grid. It is a complex network of thousands of electricity generators, transformers, switches, 160,000 miles of high voltage transmission lines, and 140 control centers. Electricity generation from 2000 to 2010 has been increasing at almost 2% per year. Present technology allows the electricity grid to grow at the necessary rate to keep up with this new demand. Electricity production has been 87% from fossil fuels which are becoming less available and more costly. Based on these figures, future availability of electricity may depend upon advances in safe nuclear energy and geothermal energy development.

OIL TRANSPORTATION

Oil that is produced in the Middle East and shipped across vast oceans to consumers in the Americas utilize the only practical and economic method to transport of oil across oceans is by large ships commonly referred to as super tankers. As of 2011, the world's two largest supertankers were the *Europe* and *Oceania*. These are double-hulled ships having a cargo capacity of 3,166,353 barrels of oil. Based on 2007 United States oil consumption of 20.6 million barrels of oil per day, it would take 6.5 super tankers to supply the United States if there were no other sources of oil such as pipelines or local production. With the exception of the pipeline, the super tanker is the most cost-effective way to move oil today. The cost of transporting oil by tanker half way around the world amounts to about 2 or 3 cents per gallon of gasoline at the pump.

CRUDE OIL PIPELINES

There are approximately 55,000 miles of crude oil trunk lines, usually 8 to 24 inches with some as large as 30 to 36 inches in diameter in the U.S. that connect oil production with regional markets. The   U.S. also has an estimated 30,000 to 40,000 miles of small gathering lines (usually 2 to 6 inches in diameter) located primarily in Texas, Oklahoma, Louisiana, and Wyoming with small systems in a number of other oil producing states. These small lines gather the oil from many wells, both onshore and offshore, and connect to larger trunk lines measuring from 8 to 24 inches in diameter.

Crude oil pipelines are the most cost effective method of transporting oil today. Cost is generally less than shipment by super tanker, but is difficult to estimate due to variations in terrain.

TRANSPORT BY RIVER BARGE

Transporting four million gallons of gasoline from a refinery in New Jersey to Gasoline Stations in Baltimore can be done by pipeline, truck, rail or river barge. Comparison of costs show that the most effective method for this particular shipment will be by river barge. You'll need a 400-foot-long double-hulled barge and a 5000-horsepower tugboat to push it. Time in transit will be longer than any of the other methods of shipment, but dost will be less than shipment by rail or truck. This cost analysis holds true for barge shipments on the Ohio and Mississippi Rivers.

Variations in shipment alternatives for oil and gasoline account for the variations in the cost of gasoline in various areas of the United States.

RAIL TRANSPORT OF OIL

A rail tank car holds about 600 barrels of oil or 25,200 gallons. It is the most cost effective method of transporting oil over land where pipe lines are not in place. Cost of rail transport of oil or gasoline is about $6 to $10 per barrel. For distances over 800 miles, rail is a cost effective method of shipment with average distances over 2000 miles.

TRUCK TRANSPORT OF OIL AND GASOLINE

When there are no pipelines, rail or barge services available to ship oil or gasoline, truck is the only form of shipment available to ship gasoline from a distribution point to a local gasoline station. Trucks can carry about 5,000 gallons of gasoline, adding about 5 cents to the retail price of a gallon of gasoline.

ABOVE GROUND STORAGE TANKS

Oil is off loaded from ships or piped direct from wells and then placed in large tanks until it is processed in the refinery. Before the oil is processed, it is stored in tanks, and after it is processed, the gasoline and other products are stored in tanks until sold.

NATURAL GAS TRANSMISSION

The natural gas transmission system can be considered analogous to the electrical power transmission system that collects, transmits and distributes electricity while the natural gas transmission system collects, transmits and distributes natural gas.

The efficient and effective movement of natural gas from producing regions to consumption regions requires an extensive and elaborate transportation system. In many instances, natural gas produced from a particular well will have to travel a great distance to reach its point of use. The transportation system for natural gas consists of a complex network of pipelines, designed to quickly and efficiently transport natural gas from its origin, to areas of high natural gas demand. Transportation of natural gas is closely linked to its storage: should the natural gas being transported not be immediately required, it can be put into storage facilities for when it is needed.

The interstate natural gas pipeline network transports processed natural gas from processing plants in producing regions to those areas with high natural gas requirements, particularly large, populated urban areas. The pipeline network extends across the entire United States and connected with Canadian pipelines.

Interstate pipelines are the 'highways' of natural gas transmission. Natural gas that is transported through interstate pipelines travels at high pressure in the pipeline, at pressures anywhere from 200 to 1500 pounds per square inch (psi). This reduces the volume of the natural gas being transported (by up to 600 times), as well as increased velocity to propel natural gas through the pipeline.

STORAGE OF ENERGY

BATTERIES

Batteries are used to store energy to power everything from an electric tooth brush to an electric airplane. This has been made possible in the 21st century by high efficiency direct current motors and batteries that are becoming more practical for transportation vehicles.

Batteries are used for energy storage in many sizes, shapes and chemical composition. Consumers are aware of disposable zinc-carbon batteries used for portable devices and lead-acid batteries (shown here) used to start their cars.

Storage of energy by means of batteries is not economically practical for large scale applications such as solar farms. However, lead-acid batteries are used for storage in critical applications such as communications where loss of primary electricity from the grid may be interrupted. A discussion of the many batteries available and their applications is included in Section Eight on Future Energy

COMPRESSED AIR STORAGE

Compressed air has been used for many purposes from blowing glass to industrial equipment. However, until recently, it has not been considered as a source of stored energy for vehicles. Tata Motors of India and other car manufacturers are developing cars that will run on a tank of compressed air. Compressed air cars will have the same environmental advantages as electric cars and can be charged in a fraction of the time to recharge an electric battery. This is discussed in greater detail in section eight.

PUMPED HYDRO

PUMPED HYDRO is a power storage facility that during low demand periods pumps water to a higher elevation so that during peak demand for electricity, the water can be used to generate electricity. One solution to

meet peak demands is to construct pumped hydroelectric facilities. Pumped Hydro is a technique that stores off peak electricity so that the energy can be used when needed at peak times. In 2009, the United States had 21.5 GW of pumped hydro storage generating capacity spread over around 40 pumped hydro facilities, all built at least two decades ago.

Pumped hydro requires a geographically suitable site that can link a high reservoir with a low reservoir and move water between the two. Off peak electricity is used to power reversible generator/pumps moving water from the low reservoir to the high one and storing its potential energy, which when needed, the water is released so that gravity will turn turbines connected to generators similar to a standard hydroelectric power generator. Approximately 70% to 85% of the electrical energy used to pump the water into the elevated reservoir is recovered.

A proposed 1.3 GW pumped hydro project, the Eagle Mountain Pumped Storage Project about 65 miles west of Palm Springs will use a closed loop system of water pumped between two existing depleted mine pits located about 14,000 feet apart, with an elevation difference between the pits of approximately 1,500 feet. Both the high and the low reservoirs are brownfield open pit mine sites that do not require any pristine valley to get flooded.

	Pumped Storage Generating Capacity	Percent of World Total Capacity
EU	38.3 GW	38.0 %
Japan	25.5 GW	24.5 %
United States	21.5 GW	20.6 %
Other Nations	18.7 GW	16.9 %
World Capacity	104.0 GW	

In 2008 world pumped storage generating capacity was 104 GW EU had 38.3 GW net capacity (36.8% of world capacity) out of a total of 140 GW of hydropower and representing 5% of total net electrical capacity in the EU. Japan had 25.5 GW net capacity (24.5% of world capacity).

In 2009 the United States had 21.5 GW of pumped storage generating capacity (20.6% of world capacity) accounting for 2.5% of baseload generating capacity.

STRATEGIC PETROLEUM RESERVES

Strategic petroleum reserves refer to crude oil inventories held by the government of a particular country, as well as private industry, for the purpose of providing economic and national security during an energy crisis. The Strategic Petroleum Reserve (SPR) is the world's largest supply of emergency crude oil. The federally-owned oil stocks are stored in huge underground salt caverns along the coastline of the Gulf of Mexico. According to the United States Energy Information Administration, approximately 4.1 billion barrels of oil are held in strategic reserves. At the moment the US Strategic Petroleum Reserve is one of the largest strategic reserves, with much of the remainder held by the other 26 members of the International Energy Agency. Dramatic worldwide drop in oil field output as suggested by some peak oil analysts, the strategic petroleum reserves are unlikely to last for more than a few months. According to a March 2001 agreement, all 28 members of the International Energy Agency must have a strategic petroleum reserve equal to 90 days of prior year's net oil imports for their respective country. This requirement is based on the havoc caused by the 1973 oil embargo by OPEC.

Decisions to withdraw crude oil from the SPR are made by the President under the authorities of the Energy Policy and Conservation Act. In the event of an energy emergency, SPR oil would be distributed by competitive sale. Prior to 2011, the SPR has been used under these circumstances only twice, first during Operation Desert Storm in 1991 and again after Hurricane Katrina in 2005.

Again, in the summer of 2011, the Obama administration tapped the SPR in order to stabilize world oil prices. U.S. Energy Secretary Steven Chu said the release of oil was in response to oil supply disruptions caused by turmoil in the Middle East and North Africa, including Libya

A few months after releasing the oil, gasoline prices at the pump were sharply lower. The 2012 presidential election was less than a year away when the oil was released giving politicians reason to claim that it was a political maneuver.

Strategic petroleum reserves can be considered as a method of taking petroleum out of the ground and at a later time putting petroleum back into the ground for reasons of economics, natural defense strategies, and political manipulation.

SECTION EIGHT
FUTURE ENERGY

Human beings have the intelligence and creative ability to build both gigantic structures and miniature computers. The earliest known structures were the Pyramids that were built over 4,000 years ago. They were built using only the muscle power and brains of man. There were no diesel powered machines to lift the heavy stones.

The industrial revolution provided man with machines to build dams, highways, bridges, tall buildings, ocean liners, and airplanes. These projects all require energy, mostly from fossil reserves. Our objectives and the objectives of the entire world are to replace depleting fossil reserves with renewable energy. Our economic wellbeing depends on our ability to develop enough renewable energy to prevent economic repercussions.

What will life be like in the future when we no longer have enough fuel to power these machines? The answer depends upon the fuel that will be available from renewable sources.

WORLD FAIR 1939 GM FUTURAMA

Over 70 years ago, people from around the world visited the General Motors Futurama exhibit at the 1939 – 1940 World Fair. They sat back on moving chairs and saw scaled models of what the highway system of the future would be like. Visions were extremely accurate as witnessed by the photo at the left compared to the interstate highway system of the 21st century.

But, little thought was given to the possibility that in the 21st century, oil to power all those cars would start to become depleted so that the cost of gasoline to power those cars would restrict the use of modern highways. Little thought was given to the possibility that international wars would be fought over oil to power gas guzzling cars of the future.

FIGHTING THE WRONG WARS

There is no doubt that we had to fight World War II. From Europe, the world was being attacked by a tyrant that intended to change the course of civilization. From Japan, the United States was being attacked by a small nation that needed resources (oil) that it intended to fight for. The situation was clear, and the course of action that the United States had to take was clear. The United States mobilized its brains and resources to stop aggression on two fronts. The United States came into the fight late, but caught up and was triumphant. After World War II, the United States saw itself as the most powerful nation in the world and took on the task of protecting the world. Protecting the world by fighting wars is expensive, especially if it is the wrong war.

We won some of the wars and lost some. In either case, those wars took an economic toll on the United States. Our politicians do not understand the consequences of spending money that we do not have when committing to fight those wars. Our politicians do not understand the economic consequences of promising to take care of the needy with money that we do not have.

Our politicians do not understand that it is cheaper to repair something rather than to send it to the junk heap and build a new one. The United States has wonderful Social Security and Medicare systems, but like all things in life, they are not perfect, so our politicians decided to fix them by taking drastic actions that will have dire consequences on our fragile economy.

Our politicians do not realize that we have an immediate energy crisis and that the magnitude of that crisis can bring down the economy of the United States and the world. Each day the situation gets worse. Actions that we are taking to get better gas mileage from our cars and to grow ethanol fuel are too little and too late.

When we were fighting World War II, the situation was easy to understand because we could see pictures of the devastation. There are no pictures that can describe our present energy situation other than to search for photos of those long gas lines back in 1973. We are losing the war on energy and most people don't even realize that we are at war.

Americans: Wake up and declare war on energy. We are fighting energy battles every day with unemployment, higher energy costs for gasoline and essential food. This is a war we can win without sending a single soldier to a foreign land, if we start fighting today.

FOSSIL FUEL DEPLETION

Fossil fuels imbedded in the earth are in solid, liquid and gas form, respectively coal, oil and natural gas. If we take them out of the earth, they will be gone forever. It is impossible to accurately know how much of each is available, and therefore, it is impossible to know how much can be extracted from the earth before reaching depletion.

Depletion of any of the fossil fuels can be defined as the point at which extraction becomes too expensive to justify further extraction.

All studies agree that depletion of fossil fuels will be in the order of oil first, then natural gas and then coal. Estimates to reach oil depletion range from 50 to 150 years, natural gas depletion from 100 to 200 years, and coal depletion better than 200 years.

Based on estimates of depletion, world oil production will reach "World Peak Oil" sometime between 2020 and 2035. Anticipation of depletion of oil is resulting in more renewable energy and alternate energy projects. Renewable energy from bioethanol and wind power has seen dramatic increases from 2000 to 2010. Alternative energy for transportation will come from conversion of coal and natural gas to gasoline for powering vehicles. Such alternative energy will cause earlier depletion of natural gas and coal as industry consumes more of them to produce gasoline for transportation vehicles.

Energy companies and vehicle manufacturing companies advertise so that they can sell more of their products, resulting in earlier depletion of fossil fuels than would have otherwise occurred. Imagine an oil company advertising not to buy its product because doing so will contribute to early depletion of oil, or an automobile manufacturer advertising not to purchase a gas guzzling SUV because it will contribute to early depletion of oil reserves.

WORLD PEAK OIL

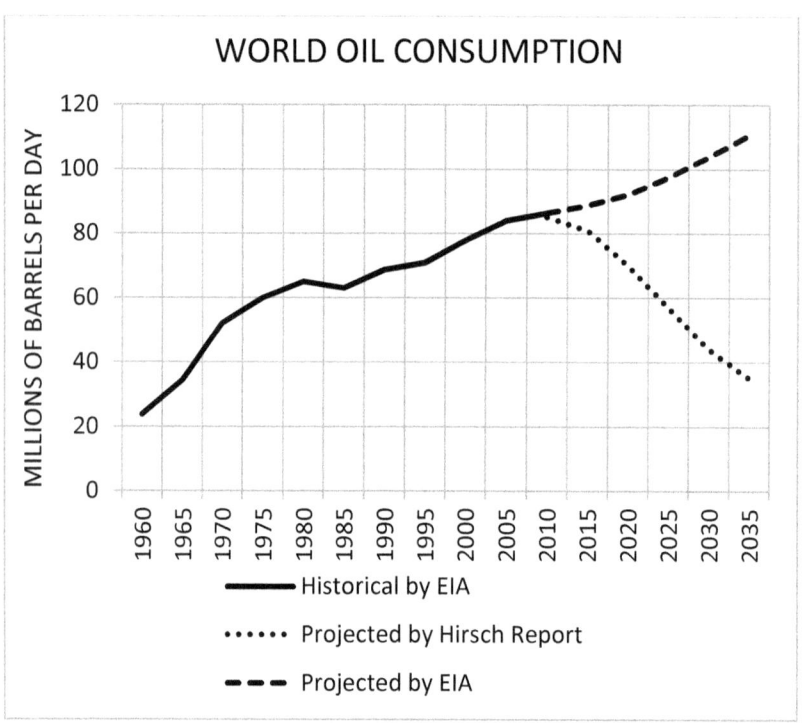

U.S. ENERGY INFORMATION ADMINISTRATION
INTERNATIONAL ENERGY STATISTICS DATABASE TABLE A5

The above chart titled "World Oil Consumption" shows a gross conflict between data presented by the United States Energy Information Administration (dashed line) and that of the Hirsch Report (dotted line).

The International Energy Agency (EIA) predicts that crude oil from currently producing fields will peak 2010. It also predicts that crude oil from "fields yet to be found" and "fields yet to be discovered" will result in a modest increase in world oil production through the year 2035. The latter was prepared at the request of the US Department of Energy. The conflict in data presents a serious question regarding "World Peak Oil". Predictions by EIA indicate that "World Peak Oil" will not occur before 2035, if ever, while the Hirsch Report shows that "World Peak Oil" will occur abut 2010. The Hirsch Report states that "The peaking of world oil production presents the U.S. and the world with an unprecedented risk management problem. As peaking is approached, liquid fuel prices and price volatility will increase dramatically, and, without timely mitigation, the economic, social, and political costs will be unprecedented. Viable options exist on both the consumption and demand sides which must be addressed by conservation, discovery of new conventional oil fields, new oil sands, and alternative sources of mobility fuel such as Ethanol and algae fuels.

WORLD OIL DEPLETION

If we assume that there will be no further oil reserves discovered and the present rate of oil consumption continues, world oil reserves will be depleted in 16,000 days or 44 years. That assumption is not realistic because new oil reserves will be discovered, but, new oil field discoveries have been declining steadily for 40 years despite extensive exploration with the most advanced technology, and most importantly, finding giant new fields is becoming ever rarer. In 2002, the world used four times more oil than was found from new sources.

Assessments of world ultimately recoverable oil reserves put world original endowment of recoverable oil at no more than about 2,400 billion barrels; the average estimate is 2,000 billion barrels. Cumulative worldwide consumption had exceeded 900 billion barrels by the end of 2003.

Peak oil is described as the point in time when the maximum rate of global petroleum extraction is reached, after which the rate of production enters terminal decline. This concept is based on the observed production rates of individual oil wells, and the combined production rate of a field of related oil wells. Peak oil is often confused with oil depletion; peak oil is the point of maximum production while depletion refers to a period of falling reserves and supply.

Peak of United States oil production has already occurred which can be seen on the graph titled US Oil Consumption and Production shown later in this section. United States oil production peaked in 1970 and has continued downward since then even though off shore drilling has increased at the greatest rate since off shore drilling began.

Peak of world oil production has been the subject of many books and articles in technical publications. All articles and publications agree that oil production will peak between 2010 and 2025. Here is the problem. When the US oil consumption necessitated importing oil, the US started importing oil from countries that had excess oil production. Question: When oil consumption here on Earth exceeds production, will we import oil from Jupiter or Mars?

Pessimistic observers, believe the high dependence of most modern industrial transport, agricultural, and industrial systems on the relative low cost and high availability of oil will in a short time will have negative implications for the global economy.

Optimistic estimations of peak production forecast the global decline will begin by 2020 or later, and assume major investments in alternatives will occur before a crisis, without requiring major changes in the lifestyle of heavily oil-consuming nations.

Of the largest 21 fields, at least 9 are in decline. In April 2006, a Saudi Aramco spokesman admitted that its mature fields are now declining at a rate of 8% per year. This information has been used to argue that Ghawar, which is the largest oil field in the world and responsible for approximately half of Saudi Arabia's oil production over the last 50 years, has peaked. The world's second largest oil field, the Burgan field in Kuwait, entered decline in November 2005.

According to a study of the largest 811 oilfields conducted in early 2008, the average rate of field decline is 4.5% per year. The IEA stated in November 2008 that an analysis of 800 oilfields showed the decline in oil production to be 6.7% a year, and that this would grow to 8.6% in 2030. Mexico announced that its giant Cantarell Field entered depletion in March 2006, due to past over-production.

Saudi Arabia's inability, as the world's largest supplier, to stabilize prices by increased production suggests that no nation or organization had the spare production capacity to lower oil prices.

The implication is that those major suppliers who had not yet peaked are operating at or near full capacity.

Conclusion must be made that peak oil will occur sometime in the next 30 years. There are several means by which the world can prepare for the inevitable decline in oil production, all of which are being addressed in different ways by different countries. The key factor in preparing for peak oil is to establish a time frame using a worst case scenario, and then declaring war on the inevitable energy crisis.

US PEAK OIL

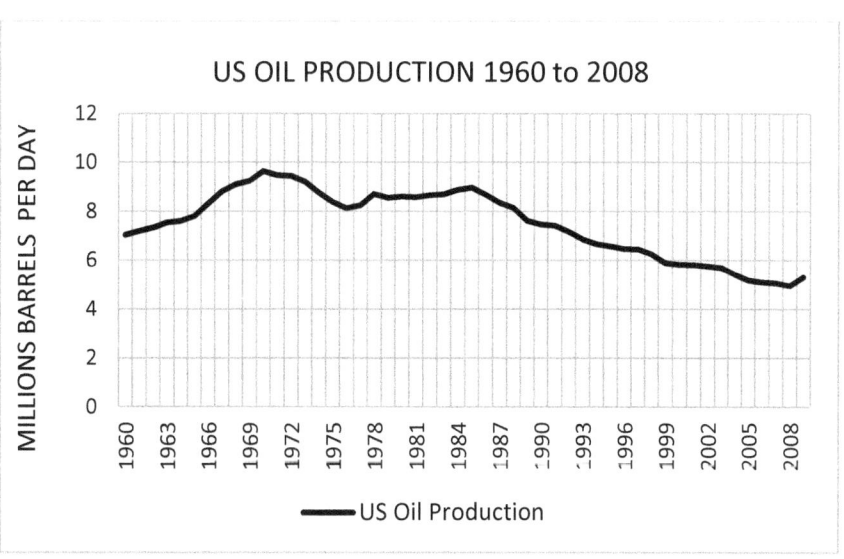

This graph shows that United States Oil Peak occurred in 1970 at about 10,200,000 barrels per day. No analysts dispute this fact The graph does not extend beyond 2008 due to lack of credible predictions for US oil production and discovery of new oil reserves.

The Hubbert peak theory states that for any given geographical area, from an individual oil-producing region to the planet as a whole, the rate of petroleum production tends to follow a bell-shaped curve. It is one of the primary theories on peak oil. Hubbert Curve predicts continuing reductions in US oil production.

Hubbert's Peak was achieved in the continental US around 1970 when oil production peaked at 10,200,000 barrels per day. Since then, it has been in a gradual decline.

UNITED STATES OIL STATISTICS

United States oil reserve life of only 4 years does not take into account new oil reserves that are constantly being discovered. However, rate of those new discoveries is diminishing and production from those new discoveries is becoming more costly.

United States oil consumption as of 2008 was 19.5 million barrels per day which is 30% of world total production.

There have been many controversies as to when world oil production will peak. However, there can be no controversy that United States oil production peaked in 1970.

Statistics show that this trend will continue until drastic steps are taken to reduce oil consumption

UNITED STATES
TRANSPORTATION OIL CONSUMPTION

2009 UNITED STATES TRANSPORTATION OIL CONSUMPTION	Million Barrels per Day	% of United States Total
Cars	4.662	24.24
Light Trucks (4 tires)	4.019	20.90
Motorcycles	.031	.16
Transit Buses	.044	.23
Intercity Buses	.015	.08
School Buses	.036	.19
Medium-Heavy Trucks	2.901	15.1
General Aviation	.109	.56
Domestic Air Carriers	.740	3.84
International Air Carriers	.187	.97
Water Freight	.046	2.41
Water Recreational	.126	.65
Pipeline	.003	.15
Rail Freight	.210	1.09

Rail Passenger Commuter	.006	.31
Rail Passenger Intercity	.004	.20
US Transportation Total	13.556	100.00
US Oil Total	19.203	
World Oil Total	82.770	

US DEPARTMENT OF ENERGY, TRANSPORTATION ENERGY DATA BOOK TABLE 1.15

Notes to the previous table:

Pipeline liquids (also referred to as natural gas liquids) include propane, butane, etc. are generally marketed as bottled gas.

Medium-Heavy Trucks are those that have more than 4 tires.

General Aviation is mostly corporate jets which have increased fuel consumption from 208 million gallons in 1970 to 1,550 million gallons in 2008, an increase of 645%. This is by far the largest increase of any sector. These corporate toys are the ultimate in inefficiency, generally take off with seating capacity of less than 25%.

Rail Passenger Commuter and Rail Passenger Intercity appear to be relatively low because they do not include trains operated on electricity.

These figures do not include consumption of renewable energy from ethanol, which was 12,000 million gallons for the year 2009.

SOME INTERESTING FACTS:

United States has 4.5% of world population, and consumes 23.2% of world oil production, which is more than the next five highest consumers of oil, China, Japan, India, Russia, and Brazil combined.

United States cars and light trucks consume 10.5% of total world oil consumption, which is more than the total oil consumption of:

      1. China or
      2. Japan and India combined or
      3. Russia, Brazil and Germany combined.

For 2009 over 45% of the crude oil consumed in the United States was in the category "cars, motorcycles, small trucks"

The US transportation sector is the largest culprit of oil consumption in the world.

2009 US total consumption	207,813 million gallons
2009 US total consumed from ethanol	12,000 million gallons
2009 US total oil production	80,000 million gallons
2009 US total from oil imports	115,813 million gallons

The United States is consuming a disproportionate share of the world's depleting fossil fuel reserves. We must take action immediately in order to prevent economic disaster, starvation and suffering in the United States and the World.

THE ENERGY BUBBLE

When this book was originally being laid out, this section discussing the energy bubble was in the section on future energy. At that time, our economic problems were thought to be a continuation of the problems caused by the Great Recession of 2008. However, economic events occurring in 2010 point to energy as a large contributor to the problems. Some indications that the energy bubble has started to burst are:

1. The Deepwater Horizon disaster showed that the demand for oil and profits are more important than the need for safety when drilling for oil. Offshore drilling is proceeding without correcting the causes of the Deepwater Horizon disaster.

2. Hydraulic fracturing to obtain natural gas is proceeding without taking proper precautions to prevent contaminating drinking water supplies.

3. Coal is being used in greater amounts without taking necessary precautions to reduce CO_2 emissions which are known to cause global warming.

The energy bubble will first be seen in the transportation industry which uses 69% of the total oil consumed in the United States. Transportation is one of the most essential parts of the economic cycle. Employees must get to work. Goods must be shipped to the factory and finished products must be shipped from the factory and the farm to the market. If the farmer does not have enough fuel to ship his crops and milk to market, they will spoil which will further exasperate the situation resulting in higher prices that the consumer can ill afford. If the energy crisis is not controlled, prices will increase to the point that consumers cannot afford goods that are essential for survival.

There will be a domino effect in all industries. Here is how some of those domino effects will result in economic disaster that could spread throughout the world.

Airline industry fortunes rise and fall on the price of fuel for aircraft. Airlines will increase the price of fares to adjust for increase in fuel prices. Passengers, already burdened by higher prices for food and gasoline will have less cash available for flying. With fewer passengers there will be less need for new aircraft resulting in cancelation of orders for new aircraft.

Aircraft manufacturers will have fewer orders for aircraft which will result in unemployment of aircraft workers, pilots, aircraft crews, and airport personnel.

Tourist industries will decline resulting in less need for hotels, motels and restaurants, causing loss of jobs in tourism which is an essential part of the national income for many small nations.

Automobile industry sales were adversely affected during the Great Recession of 2008. This is starting to occur again with many potential automobile purchasers waiting to see what will be available in hybrid cars and all electric cars.

This domino effect will continue to adversely affect the economy until our demand for oil is reduced to the level of supply.

Have we learned from our past experiences that bubbles will burst when allowed to go unchecked? PROBABLY NOT. Will our political and industrial leaders prevent the looming energy bubble from bursting? PROBABLY NOT.

Action that can be taken will not be pleasant, but it will better than the alternative of taking no action and allowing our economy to fall into economic recession.

Action such as rationing will be unpleasant for all, but taking no action will result in a recession that will cause large-spread starvation and suffering to a large percentage of the population. Not the wealthy. The wealthy will be able to purchase gasoline for their gas guzzling cars, yachts and private airplanes.

 The poor and middle class will see the inequities in our system. They will understand why we got into the situation. They will see that the government is not capable of handling the situation. They will see how the 21st century spawned revolts in depressed nations. Will Americans revolt? The question is whether the revolts will be peaceful or bloody.

In many respects, the Great Recession of 2008 is analogous to the Great Depression of 1929.

In the Great Depression of 1929, consumers, many of whom had suffered severe losses in the stock market the previous year, cut back their expenditures. Beginning in 1930, a severe drought ravaged the agricultural heartland of the USA contributing to the Depression

In the Great Recession of 2008, consumers who saw the values of their homes deteriorate as the result of the housing meltdown cut back their expenditures. The looming oil depletion further contributed and extended the Great Recession of 2008.

It took us 20 years to recover from the Great Depression of 1929. Recovery was not due to the efforts of our political leaders, but rather because a foreign dictator forced us into World War II, necessitating us to unite in a manner that our government was not capable of accomplish otherwise.

It could take many years to recover from the Great Recession of 2008. It will not come from our political leaders who should take action to ration our depleting oil reserves, but rather, recovery will come from scientists who find renewable energy to replace depleting oil reserves, and from entrepreneurs who invest in businesses that deal with the depleting oil reserves.

Can our political leaders comprehend that we must declare war on energy? PROBABLY NOT. Can our political leaders unite both the rich and the poor to fight the war on energy? PROBABLY NOT.

TRANSPORTATION VEHICLES

Transportation vehicles either carry their own source of energy, or connect to electric lines for propulsion. Although some experimental vehicles designed by engineering students utilize solar panels to charge batteries or sails to capture the wind, such vehicles are considered to be impractical.

All transportation vehicles have some sort of engine to propel them, usually an internal combustion engine or an electric motor.

With a looming energy crisis just around the corner, we are focusing on the efficiency of everything from light bulbs to automobiles. The old incandescent light bulb was very inefficient as can be seen, or rather felt, by touching the bulb. The heat that was felt came from electrical energy heating the bulb and the surrounding air rather than producing light. In the case of the light bulb, technology is on the right track providing us with alternatives to the inefficient light bulb.

Most people who drive a car have felt the heat being ejected from the radiator which is wasted energy not being used to propel the car. Modern cars that claim 30 or 50 miles per gallon still require a radiator that sends heat from the engine into the atmosphere. They are very inefficient. Over the past 20 years, technology has improved the efficiency of the internal combustion engine, but has reached the point where better efficiency is no longer practical unless the engine is replaced by the Atkinson cycle engine presently used in most hybrid cars. More details are discussed in the paragraphs under hybrid vehicles.

There are two types of transportation vehicles differentiated by the type of energy used to propel them. Some vehicles such as large commercial aircraft and ocean going ships must carry their own fuel, referred to as mobility fuels. When fuel must be carried onboard the vehicle such as an airplane, it must be in a state that is safe and practical.

Aircraft cannot carry liquefied natural gas (LNG) because the high pressure tanks would be too heavy and susceptible to rupture in the event of a hard landing. The 21st century saw new technology which allowed electric aircraft to sustain flight for 2 hours. Cessna Aircraft developed an electric airplane that could be used for training pilots but due to short duration would be impractical for commercial aircraft.

Automobiles (including motorcycles) primarily use gasoline derived from oil. As oil prices increase due to shortages, automobiles are slowly switching to biofuels, liquefied natural gas, ethanol, and most significant, electricity. Electric vehicles require high capacity batteries which use Lithium and high power electric motors which require Cobalt in their magnets. Availability of Lithium and Cobalt remain a key factor in the development of electric vehicles.

GASOLINE VEHICLES

Cadillac Fleetwood shown here is typical of the gas guzzling cars of the 1970s when gasoline was less than a dollar a gallon. When purchasing a car like the Cadillac, the last thought was how many miles per gallon the car would get, until the 1973 oil embargo caused long lines at the gas pumps and for the first time, drivers gave thought to our dependence on foreign oil.

Prior to the 1973 oil embargo, drivers were aware that there was more power under the hood than would ever be necessary, but the manufacturers were concerned that if they put smaller engines in their cars it might affect sales, and the oil companies wanted the larger engines because the gas guzzlers would consume more gasoline.

Most drivers have never "put the pedal to the metal", and some are actually afraid what would happen if they unleashed all that power.

As gasoline prices increased, drivers complained about the high price for gasoline, but it did not deter them from buying large cars and SUVs. Contrary to the United States, most countries that depend upon foreign oil knew that they had to take action to reduce that dependency. Countries around the world placed a tax on transportation fuels which necessitated conservation, generally by driving smaller more efficient vehicles. The United States government did nothing. The oil companies said nothing about our depleting oil reserves, and the automobile manufacturers encouraged drivers to buy large inefficient SUVs. Now our politicians finally acknowledge that there is an energy problem and that we are dependent on our enemies to provide us with enough oil to prevent an economic collapse. They are addressing the situation in the best way they know how, but politicians are limited in their understanding of how our fossil fuels are becoming depleted. Our automobile executives, who have firsthand knowledge of our depleting oil reserves, continue to promote large, powerful, expensive, prestigious gas guzzling cars because they are concerned only with the profits such cars will bring to their bottom line.

DIESEL VEHICLES

Diesel engines produce more power per gallon of fuel than gasoline engines and are available with higher power ratings than gasoline engines which is necessary for pulling heavy loads. In 2009, medium-heavy trucks, usually with diesel engines, consumed 44,000 gallons of fuel. This was about 33% of the fuel consumed by all cars and light trucks. It was almost twice the fuel used by all United States air carriers.

Due to the high cost of diesel fuel, goods shipped long distances in semi-trailers (18 wheelers) are more frequently utilizing rail to reduce shipping costs. However, there are practical limitations when transporting semi-trailers by rail for short distances such as the location of loading terminals plus delivery of the semi-trailer to the terminal.

COMPRESSED NATURAL GAS VEHICLES

Natural gas has long been considered an alternative fuel for the transportation sector since the 1930's.

There are currently 150,000 Natural Gas Vehicles (NGVs) on the road in the United States today, and more than 5 million NGVs worldwide. In fact, the transportation sector accounts for 3 percent of all natural gas used in the United States. Most natural gas vehicles operate using compressed natural gas (CNG). This compressed gas is stored in similar fashion to a car's gasoline tank.

In addition to being environmentally friendly, natural gas is an economic alternative to gasoline and other transportation fuels. Natural gas vehicles have been around 30 percent cheaper than gasoline vehicles to refuel, and in many cases the maintenance costs for NGVs is lower than traditional gasoline vehicles

In order to reduce consumption of fuel derived from oil such as gasoline and diesel, the government is allowing tax credits of up to $7,500 toward the purchase of consumer compressed natural gas and liquefied natural gas vehicles, and as much as $64,000 for heavier grade commercial trucks.

Worldwide, there were 12.7 million natural gas vehicles in 2010, led by Pakistan with 2.7 million. As of 2009, the U.S. had a fleet of 114,270 compressed natural gas (CNG) vehicles, mostly buses. The number of NGV refueling stations has also increased, to 18,202 worldwide as of 2010, up 10.2% from the previous year.

LIQUEFIED NATURAL GAS VEHICLES

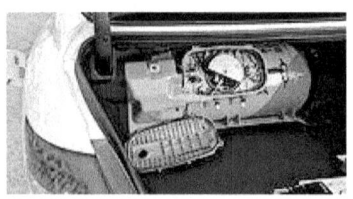

A small LNG tank can be placed in the trunk of a car and filled at a specially equipped LNG fueling station. Note the small size of the LNG tank which is made possible by the higher density of LNG.

LNG will not be available at regular gasoline service stations in the near future because of the high cost of installing the special equipment which costs about $1 million. As of 2009 there were only 3,176 vehicles in the United States equipped for liquefied natural gas (LNG).

As a liquid, LNG is not explosive. LNG vapor will only explode in an enclosed space within the flammable range of 5-15 percent. LNG is produced both worldwide and domestically at a relatively low cost and is cleaner burning than diesel fuel.

Since LNG has a higher storage density, it is a more viable alternative to diesel fuel than compressed natural gas for heavy-duty vehicle applications. Because of LNG's increased driving range, it is being evaluated in heavy-duty vehicles including railroads on a trial basis.

U. S. NATURAL GAS RESERVES CONTROVERSY

Natural gas reserves and methods of extraction have become controversial starting about January 2009 when some estimates of "Natural Gas Reserves" started to increase dramatically. The Natural Gas Industry would like to sell more of their product and one way to sell a product is to have consumers believe that there is a large reserve of their product. The question remains as to why controversial methods of extraction such as hydraulic fracturing must be used if there are large reserves available. Why not tap reserves that are more readily accessible?

THE ENERGY INFORMATION ADMINISTRATION report dated January 1, 2009 estimated U.S. Natural Gas Reserves at 244 trillion cubic feet. A recent report by the Energy Department dated April, 2011 dramatically increased reserves to 2587 trillion cubic feet mostly from increased recoverable shale reserves to be technically recoverable. Assuming production remains constant at 2011 rates, reserve life would be extended to 40 years.

U.S. Natural Gas Reserves estimate 2009	Trillion cubic feet
Undiscovered Reserves	441
Inferred Reserves	1032
Unconventional Gas	493
Total Unproved	2349
Proven Reserves	238
Total Proven and Unproved	2587

THE ENERGY INFORMATION ADMINISTRATION
ANNUAL ENERGY OUTLOOK 2010

Based on the above tabulation, actual U.S. Natural Gas Reserves are somewhere between 238 and 2587 trillion cubic feet. These figures do not leave the reader with a warm confident feeling that natural gas reserves are the answer to our depleting oil reserves.

THE NATIONAL PETROLEUM COUNCIL in its 2007 report, "Facing the Hard Truths about Energy" estimated total traditional natural gas resources to total 1,451 trillion cubic feet.

U.S. Natural Gas Reserves estimate 2007	Trillion cubic feet
US Onshore	764
US Offshore	384
Alaska	303
Total traditional gas resources	1,451

NATIONAL PETROLEUM COUNCIL
FACING THE HARD TRUTHS ABOUT ENERGY 2007

THE POTENTIAL GAS COMMITTEE, May, 2011 estimated United States reserves to be 2,170 trillion cubic feet attributed to expanded shale gas reserves. Assuming production remains constant; United States reserve life would be extended to 103 years. Shale gas represents approximately one third of this estimate.

U.S. Natural Gas Reserves estimate ending 2010	Trillion cubic feet
Probable resources (current fields)	537
Possible resources (new fields)	688
Speculative resources (frontiers)	518
Total traditional gas resources	1,739

POTENTIAL GAS COMMITTEE, APRIL 2010

THE CENTRAL INTELLIGENCE AGENCY, The World Factbook, January 2010 estimated United States natural gas reserves at 244 trillion cubic feet. These are proven reserves without including estimates for unproven reserves.

LOS ANGELES TIMES May 17, 2011 - The United States does not have a decades-long supply of inexpensive, locally sourced natural gas, according to a new report commissioned by the Post Carbon Institute, a nonprofit think tank that examines issues related to the economy, energy and the environment. The report, titled "Will Natural Gas Fuel America in the 21st Century?" is a challenge to the commonly cited projection that domestic natural gas can meet U.S. demand for more than 100 years. The report takes the position that natural gas reserves have been greatly overstated. Some geologists estimate that there is only a 50 year supply of easily accessible domestic natural gas.

The article includes the position that environmental groups have raised concerns about groundwater contamination and the improper treatment of the fluids used in hydraulic fracturing, or fracking, to extract natural gas from shale. The U.S. Environmental Protection Agency is investigating hydrofracking and will release its study of the practice in late 2012.

For at least the next 20 years, consumption of natural gas will increase at a greater rate than gasoline and coal. Greater consumption of natural gas can be attributed to environmental concerns, cost, availability, new uses, and ease of transporting. New uses include electric generation, compressed natural gas vehicles, natural gas to gasoline conversion.

Depletion of natural gas in the United States can be considered analogous to the depletion of oil characterized by deeper wells, wells in remote areas, more risk to the environment, less production per well and increased depletion of existing wells.

Whether world natural gas reserves last for 50 years or 100 years is in question, but what is not in question is the fact that eventually natural gas will be depleted because it is a fossil fuel that is not renewable. Depletion of natural gas is defined similar to oil, which is the time when it will no longer be economically practical to produce more natural gas.

HYDRAULIC FRACTURING

Extraction of Natural Gas has become controversial due to the negative effects of hydraulic fracturing, often called Fracking, which is the process of initiating and subsequently propagating a fracture in a rock layer, by means of a pressurized fluid, in order to release petroleum, natural gas or other substances for extraction.

The energy from the injection of a highly pressurized fluid, such as water, creates new channels in the rock which can increase the extraction rates of fossil fuels. Energy producers such as Chesapeake Energy defend hydraulic fracturing by stating that hydraulic fracturing, combined with sophisticated horizontal drilling, provides access to extraordinary amounts of deep shale natural gas and oil from across the United States.

The practice of hydraulic fracturing has come under scrutiny internationally due to concerns about environmental and health safety, and has been suspended, banned or a moratorium is in place in several countries including France, Australia, New Zealand, South Africa and the United Kingdom.

Fracturing fluids that are injected at high pressure into the ground contain about 90% water plus biocides and certain petroleum products which are known to be hazardous chemicals that may cause health risks that range from rashes to cancer. Some chemicals are identified as carcinogens, and some chemicals found injected into the earth identify as endocrine disruptors, which interrupt hormones and glands in the body that control development, growth, reproduction and behavior in animals and humans. About 750 chemicals have been listed as additives for hydraulic fracturing in a report to the US Congress in 2011.

In the US, Congress exempted fracturing fluid from regulations of the Safe Drinking Water Act in 2005. Reasons for this exemption are political and may never be disclosed to the public.

Hydraulic fracturing has become a triple whammy because contaminated fluids can enter local drinking water from three sources.

> 1. Pipeline liquids such as Propane, Butane, and other toxic chemicals inherent in raw natural gas can enter the local drinking water through cracks created by hydraulic fracturing.
> 2. Fracking chemicals which are added to the hydraulic water, many of which are toxic, can enter the local drinking water.
> 3. Hydraulic fracturing requires large amounts of water which is sometimes taken partially treated from the municipal water treatment plant, rather than expensive fresh water. This water and can still contain toxic chemicals which may enter the local drinking water.

The Natural Gas industry has been put under scrutiny because of bad publicity about hydraulic fracturing to produce natural gas. In September and October of 2011, Exxon Mobile produced television commercials claiming that hydraulic fracturing is safe. If it really was safe, it would not be necessary for Exxon Mobil to defend its operations.

Some scientists and environmental groups consider Fracking as an accident (catastrophy) waiting to happen.

DOCUMENTARY GASLAND

A "NOW on PBS" episode aired in March 2010 introduced the documentary film "GasLand" which discusses the negative effects of Fracking to obtain Natural Gas, from poisoned water sources to kitchen sinks that burst into flames and health problems in the areas of natural gas Fracking. It interviews lobbyists for the natural gas industry who reluctantly disclose the chemicals, many of which are poisonous, that are mixed with the water used for Fracking, which eventually work their way into ground water used for domestic use including drinking water. This documentary film is available on the internet titled "GasLand".

LIQUEFIED PETROLEUM GAS

LPG, or liquefied petroleum gas, is a mixture of gasses that stay liquid at a practical temperature and pressure. It is considerably cheaper than natural gas not only for pumps and storage but also the fuel itself. However, it has drawbacks because it is heavier than air and tends to collect below a vehicle or tank during leaks, which would allow it to ignite explosively upon ignition. Other draw-backs include up to 20 minutes refueling time for a sedan,

It is rarely used in developed countries as a transport fuel due to above mentioned drawbacks.

HYBRID VEHICLES

Hybrid vehicles are becoming popular because they get better gas mileage than an equivalent vehicle with a standard internal combustion engine vehicle. Why? The standard four cycle "Otto" cycle internal combustion engine with efficiency as low as 15% has been replaced by an Atkinson cycle engine which has better fuel efficiency when operated at a specific speed and load. Compared to an Otto cycle engine producing 250 horsepower, an Atkinson cycle engine of similar displacement may produce only 140 horsepower.

While a modified Otto cycle engine using the Atkinson cycle provides good fuel economy, it is at the expense of a lower power-per-displacement as compared to a traditional Otto four-stroke engine. If demand for more power is intermittent, the power of the engine can be supplemented by an electric motor. This forms the basis of an Atkinson cycle-based hybrid electric drivetrain. Eelectric motors can be used independently of, or in combination with, the Atkinson cycle engine, to provide the most efficient means of producing the desired power for acceleration.

Hybrid cars can have all or as little as only one of the following features:

Regenerative braking
Atkinson cycle engine
Electric propulsion motor and battery
Generator to charge the propulsion battery
Electronics control package
Engine is off when vehicle stops

By 2012, practically every auto manufacturer and every model will be available as a hybrid, making the purchase of a hybrid vehicle a challenge in higher mathematics. Before the hybrids came on the market, purchasing a car was rather simple because most cars had features that were mandated by government safety requirements, and many other features that were expected in all cars except the very low priced sub-compact cars. The only differences between cars were engine horsepower, number of cylinders, and number of doors, bells and whistles, physical dimensions, and of course price.

There is a long list of hybrid cars with practically every car manufacturer producing or planning to produce a hybrid version of every model. If you do your homework and select the right hybrid, purchasing now may be a good idea because;

The cost of hybrid cars may not be coming down in the immediate future because there is an impending increase in the costs of many rare materials used in the manufacture of hybrid cars. For example, the rare earth element dysprosium is required to fabricate many of the advanced electric motors and battery systems in hybrid propulsion systems. Neodymium is another rare earth metal which is a crucial ingredient in high-strength magnets that are found in permanent magnet electric motors. Nearly all the rare earth elements in the world come from China, and many analysts believe that an overall increase in Chinese electronics manufacturing will consume this entire supply in the near future. In addition, export quotas on Chinese rare earth elements have resulted in an unknown amount available for export.

A few non-Chinese sources such as the advanced Hoidas Lake project in northern Canada as well as Mount Weld in Australia are currently under development, however, the economic barriers to entry are high and require years to go online.

Before purchasing a hybrid, read the following paragraph. It may save you a lot of money.

THE HYBRID HOAX

Some hybrid cars such as the Toyota Prius have achieved recognition by their fuel efficiency, performance and quiet operation. Success results in imitation with auto manufacturers jumping on the bandwagon eager to sell their cars by cashing in on the word "hybrid".

The consumer must be aware that there is no industry standard for a hybrid car. Some hybrid cars on the market may not include an essential feature to achieve fuel economy, which is an internal combustion engine operating on the Atkinson cycle. Therefore, the consumer must be aware that the word "hybrid" in the name does not guarantee fuel efficiency, performance and quiet operation.

When purchasing a car with hybrid in the name, the buyer must be assured that he is purchasing a car that will justify the initial extra cost over a similar car that is not a hybrid. To do this:

First, calculate the cost per mile for an equivalent non-hybrid car by dividing the cost of a gallon of gasoline (example use $4.00) by the miles per gallon MPG (example use Honda Civic 30 MPG) to determine the cost of gasoline per mile, $0.133 per mile.

Second, calculate the cost per mile for the Hybrid car the same way by dividing the cost of a gallon of gasoline (same example $4.00) by the miles per gallon MPG (example use Honda Civic Hybrid 44 MPG) to determine the cost of gasoline per mile, $0.091 per mile.

Third, calculate the difference in cost per mile between a non-hybrid and a hybrid by subtracting $0.133 minus $0.091 to obtain a cost savings of $0.042 per mile.

Fourth, calculate the additional price for the hybrid which is $24,000 minus $15,500 for the standard version to obtain the premium paid for the hybrid of $8,500

Fifth, calculate the number miles necessary to drive the hybrid to recuperate the additional cost of the hybrid by dividing $8,500 by $0.042 to obtain 202,380 miles necessary to drive the hybrid to achieve break even.

Let's try the calculations on a 2011 Cadillac Escalade:

Cadillac Escalade 2011	$63,455	16 MPG
Cadillac Escalade Hybrid 2011	$74,135	21 MPG

Calculates to 178,000 miles to break even.

PLUG-IN HYBRID VEHICLES

The shopping center is only 7 miles from home. Driving the Honda Civic Hybrid will consume between 0 and ¼ gallon of gasoline for the round trip depending on how much charge is in the battery. Not much can be done about the amount of charge in the battery unless the car has plug-in provisions.

The plug-in feature of hybrid cars is intended to be used at the driver's residence to charge the battery overnight and then use the car the next day to go to the local shopping center without using a single drop of gasoline.

To charge the plug-in hybrid, you simply plug it into any standard household outlet with a dedicated 15-amp circuit to charge the battery in three hours. Cost of electricity to charge the battery will be about $1.00 depending on several variables such as the local cost of electricity and the efficiency of the charger. When the plug-in hybrid is used for commutes within the range of a battery charge, it is possible to avoid filling the car with gasoline.

Plug-in hybrids are rated by the miles on a full charge plus the miles per gallon in the hybrid mode. If the battery runs out of charge while driving, the vehicle automatically starts the engine and the trip is finished in the hybrid mode.

Charging stations at shopping malls and on highways are becoming more common and will provide an incentive to shopping at a particular shop and at the same time a green alternative to stopping at the gas station to fill up.

Legislators have realized the importance of green vehicles by giving tax credits for plug-in hybrids and electric cars. A combination of local and national credits—up to $7,500 at the federal level, plus a $2,000 credit for charging equipment installation, plus state-based incentives (of $5,000 in California) represent so far the largest bundle of incentives ever for private purchasers of green electric-drive vehicles.

Plug-in hybrid vehicles announced as of 2012 include Toyota Prius, Chevy Volt, Ford Escape, Ford Fusion, Volvo V70, Suzuki Swift Plug-in, Audi Ai e-tron and Mitsubishi PX-Miev. Before purchasing a plug-in hybrid, cost of eventually replacing the battery, cost of electricity to charge the battery, range per charge and time per charge must be taken into consideration.

ELECTRIC VEHICLES

Take a ride in an electric golf cart or an electric car. Notice how little heat is expelled from the electric motor. Don't ride too long in that golf cart or electric car because the battery will run down, which brings up the next point.

The ideal mode of transportation vehicle would be an electric car which does not pollute the atmosphere, having its energy stored in high technology batteries that can propel the car for 500 miles, and being able to rapidly charge the battery from renewable energy sources such as wind or hydroelectric sources. We have two out of three, the electric car and renewable energy to charge the battery. Scientists and engineers are working on practical batteries with better capacity, but until that is achieved, the public has shown a willingness to accept electric cars with limited range.

The wild card in the electric vehicle future is development of batteries that are reasonably priced with long life that can propel a car for 500 miles.

The most common pure electric vehicle is the golf cart. It uses low cost lead-acid batteries to propel the cart for an entire day at a speed of about 5 miles per hour. They have become the standard method of chasing after that little white ball. Some private communities permit their use on roads alongside vehicles designed for highway use. They are virtually maintenance free and operate for pennies per mile.

Electricity is one of the oldest automobile propulsion methods still in use today; it predates the invention of the internal combustion engines of Diesel's and Benz's Otto cycle-engines by several decades.

In 1828, a Hungarian invented an early type of electric motor, and created a tiny model car powered by his new motor. In 1834,

Vermont blacksmith Thomas Davenport, the inventor of the first American DC electrical motor, installed his motor in a small model car, which he operated on a short circular electrified track. In 1838, Scotsman Robert Davidson built an electric locomotive that attained a speed of 4 mph. Between 1832 and 1839, Robert Anderson of Scotland invented a crude electrical carriage. A patent for the use of rails as conductors of electric current was granted in England in 1840, and similar patents were issued to Lilley and Colten in the United States in 1847. Rechargeable batteries that provided a viable means for storing electricity on board a vehicle did not come into being until the 1840s.

The General Motors EV1 was an electric car produced and leased by the General Motors Corporation from 1996 to 1999. It was the first mass-produced electric vehicle from a major U.S. automaker, and the first GM car designed to be an electric vehicle from the outset.

While customer reaction to the EV1 was positive, GM discontinued production. The EV1's discontinuation remains controversial, with electric car enthusiasts, environmental interest groups and former EV1 lessees accusing GM of self-sabotaging its electric car program to avoid potential losses in spare parts sales, while also blaming the oil industry for conspiring to keep electric cars off the road.

The demise of the EV1 is the subject of a 2006 documentary film entitled *Who Killed the Electric Car?* Much of the film accounts for GM's efforts to demonstrate to California that there was no demand for their product and then to reclaim every last EV1 and dispose of them. GM responded to the film's claims, laying out several reasons why the EV1 was not commercially viable at the time. The opposing theory discussed in the documentary is that the EV1 program was eliminated because it threatened the oil industry.

The Tesla Roadster shows that an electric car with performance equivalent to an internal combustion engine car can be built if cost is no object. Although an engine dependent on combustion will never attain the efficiency of the Tesla electric powertrain, the high cost will never justify the savings in fuel cost.

The Tesla Roadster, which began shipping to customers in 2008, uses Lli-Ion batteries to achieve up to 244 miles per charge while also capable of going from 0 to 60 miles per hour in under 4 seconds. High cost of the Tesla is attributed to the large amount of expensive batteries required to achieve the range.

Let's get back to the real world of affordable electric cars having excellent performance and range.

Nissan Leaf is one of the first mass produced cars available in the United States, the size of an intermediate size sedan, good acceleration, range of 100 miles per charge, speeds of 90 miles per hour, federal tax incentives, and a price that can be justified in fuel savings.

Range of the Leaf can vary from 138 miles under ideal conditions to as little as 62 miles in stop and go traffic on a cold day at 14 degrees Fahrenheit with the heater using a lot of the energy heating the car. This brings up the fact that back in the 1950s, cars were sold with a heater as an optional accessory, and today, many cars still sell air conditioning as an optional accessory. Considering the fact that drivers are not yet ready to make sacrifices because of the oil crisis and still want full climate controls, ratings are based on full use of climate controls. That's the bad news. The good news is that electric cars will provide cost savings on gasoline and possibly save the day if there is another oil embargo.

FORD FOCUS ELECTRIC - The Focus E electric car is Ford's big splash into the still-small mainstream EV market. The new-for-2011 redesign of the Focus has already garnered much positive attention and is now seen as one of the most competitive compact cars available. The E swaps the economical 4-cylinder for a battery pack, single speed transmission and electric motor. Mileage hasn't been released, but Ford claims a 100-mile maximum range for approximately $30,000

BYD ELECTRIC CARS - Chinese brand BYD (Build Your Dreams) is bullish on its prospects in the United States. With over 500,000 vehicles sold in China in 2010, it's one of China's top brands and a leading global battery producer. BYD features a lithium-ion-powered 200-mile-range and an 87 mph top speed. Initial tests suggest "sluggish" acceleration. Pricing has not yet been announced.

CODA AUTOMOTIVE, a southern California automaker announced plans to bring a new electric car to the US from China in 2011. Re-engineered with a lithium ion battery, the Coda sedan promises a driving range of 120 miles. The scrappy California company may be the first start-up to offer a practical and affordable electric car to mainstream buyers. Price $44,900

THE MITSUBISHI I (formerly i-MiEV) is one of the front-runners in the race for a mass-market EV. It claims a max speed of 80 mph and a range of about 75 miles. But the Mitsubishi i's small size and modest electric drivetrain will keep it off some shopper's list. Base-level price starts at $29,900, but quickly jumps up with options

ELECTRIC VEHICLE HOAX

There is no dispute that the initial cost of an electric car is greater than an equivalent gasoline powered car. Most of that additional cost is for the expensive Lithium battery.

If the higher cost and limited range can be tolerated, then two things must be considered to determine whether an electric car is a good investment. First, that expensive Lithium battery has a limited life based on the number times it is recharged. Electric car manufacturers have different warrantees on the life of the battery, and different prices to replace the battery at the end of its life. Best way to determine the cost per mile for that battery is to ask the car dealer to put in writing. Good luck.

Put the depreciation of the battery aside, here is how to compute the fuel savings over the life of the car:

First, calculate the cost per mile for an equivalent, non-electric car by dividing the cost of a gallon of gasoline (example use $4.00) by the miles per gallon MPG (example use Honda Civic 30 MPG) to determine the cost of gasoline per mile, $0.133 per mile.

Second, calculate the cost of a full electric charge by the full electric capacity of the battery (24 KWh for the Nissan Leaf) by the cost of a KWh of electricity (national average of $.1133) and assume a charging efficiency of 80% to obtain a cost of $3.40 for a full charge.

Third, calculate the cost per mile to drive the electric car $3.40 (for a full charge) by the miles driven on a full charge (100 miles) to obtain a cost of $0.034 per mile.

Note: General Motors claims that their Volt will cost "less than purchasing a cup of your favorite coffee" to recharge. They claim the Volt should cost less than $0.02 per mile to drive on electricity, compared with $0.12 per mile on gasoline at a price of $3.60 a gallon.

Fourth, calculate the difference in cost per mile between a non-electric and an electric by taking the cost per mile of a standard non-electric of $0.133 minus and subtracting the cost per mile of an electric of $0.034 to obtain a cost savings of $0.10 per mile.

Fifth, calculate the additional price for the electric car which is $25,000 (after incentives) minus $15,500 for an equivalent non-electric (Honda Civic) to obtain the premium paid for the electric car of $9,500.

Sixth, calculate the number miles necessary to drive the electric car to recuperate the additional cost of purchasing the electric car by dividing $9,500 by the cost saving of $0.10 to obtain 95,000 miles to achieve break even. Keep in mind that this may be the time that the Lithium battery will need replacement.

ELECTRIC AIRPLANES

 Flying has fascinated man since he observed birds flying effortless in the sky. Man has flown aircraft with hot air, lighter than air filled with helium, aircraft with wings, helicopters without wings, aircraft with engines, aircraft without engines, aircraft powered by human energy, aircraft powered by air currents, aircraft powered by solar energy, aircraft powered by batteries, and aircraft without a pilot. The list goes on forever.

In 1990 the solar powered airplane Sunseeker successfully flew across the USA. It used a small battery charged by solar cells on the wing to drive a propeller for takeoff, and then flew on direct solar power and took advantage of soaring conditions when possible

In 2001, a lightweight unmanned solar-powered aircraft, the Helios flew to an official altitude record of 96,800 feet.

In 2010, Cessna announced that they were planning to build a 100% electric aircraft. Together with Bye Energy, Inc. they intend to build a proof-of-concept 172 Skyhawk. According to reports, the 4-seater could have as much as a four-hour flight time which would be ideal for a flight trainer.

Today, radio controlled model airplanes powered by electric motors and batteries are a common recreational hobby around the world.

The future of small electric airplanes will depend upon development of high capacity, safe, reasonably priced batteries.

THE RACE IS ON

Gasoline (and diesel) fueled vehicles are by far the most popular vehicles and will continue to predominate as long as fuel remains below $20.00 per gallon and the cost of Lithium-ion batteries remains high. While gasoline in the United States remains around $4.00 per gallon, there will be little incentive to switch to electric, partially electric, or natural gas. The race will change if the government decides to protect depleting oil resources by imposing higher taxes on fossil fuels, or rationing fossil fuels.

Hybrid vehicles will increase in popularity because they can utilize a smaller, less costly battery than a plug-in hybrid or an all-electric vehicle.

Plug-in hybrid vehicles will increase in popularity when the cost of Lithium-ion batteries is reduced to a competitive price. When plug-in hybrids were first introduced, the cost of the battery was almost half the selling price of the car. Savings in gasoline could not justify the initial cost of a plug-in hybrid car.

Natural gas is used much less in the United States than in the rest of the world. This is probably due to the low cost of gasoline and the lack of infrastructure to support natural gas. Fluctuations in the price of natural gas and the uncertainty of natural gas reserves contribute to lack of acceptability of natural gas in the US.

Electric vehicles will inevitably predominate the highways of the United States when oil (and natural gas) approach depletion and electric vehicle batteries are better perfected.

The wild card in the transportation game will be the future development of batteries.

BATTERY TECHNOLOGY

ZINC-CARBON BATTERIES come in many shapes and sizes, some small enough for a hearing aid and some large enough for portable radios and flashlights. Normally they are not rechargeable. Shown here in order of size are AAA, AA, C, D and 9 volt batteries. It is estimated that the AAA battery shown here will propel an average size automobile about 12 feet at 1 mph before exhaustion.

NICKEL–ZINC RECHARGEABLE BATTERIES are the next step up, often used in cordless power tools, cordless telephone, digital cameras, battery operated lawn and garden tools, professional photography, flashlights, electric bike, and light electric vehicles.

LEAD-ACID BATTERIES have been used in golf carts and electric vehicles for many years because of their price and availability. Their drawback is a low specific energy (watt-hours per pound). In order for electric cars to compete with gasoline cars it was necessary to have batteries with reasonable size and enough energy to drive a vehicle at least 100 miles between charges. This requirement led to the introduction of the Lithium-ion battery for applications requiring large amounts of energy in a small package.

Because of their price, lead-acid batteries are used in gasoline powered vehicles for starting the engine, golf carts, some electric cars, and uninterruptible power supplies.

Battery technology advances will allow car manufacturers to replace gas guzzling internal combustion engines with energy efficient electric motors powered by batteries charged by renewable energy from hydroelectric, wind power and eventually by clean, safe nuclear energy.

	Watt Hours per kg	Status
Lithium-ion	100-250	In production
Lithium Iron Phosphate	90-110	In Production
Lithium-sulphur	250	development
Lithium-titanate	100-250	development
Sodium-ion	400	development
Electric double-layer capacitor	30-85	development
Lead-acid	30-40	In production

LITHIUM-ION BATTERIES are used in most hybrid, plug-in hybrid and electric vehicles. Due to their high specific energy (watt-hours per pound) they are used in most vehicles designed for highway use. They are costly due to their design and the availability of materials used in their construction. Many battery manufacturers, private organizations and the United States Department of Energy have programs to develop less costly, longer life, and smaller batteries for the rapidly growing electric vehicle market.

Over the next five years, United States manufacturers hope to catch up on the lead of Asian manufacturers that now supply a major portion of lithium-ion batteries used in hybrid and electric cars.

Advanced battery research and development is making progress towards increased energy, power, and lifetime, with improved safety at a reduced cost. LIB are one of the most popular types of rechargeable battery for portable electronics, with one of the best energy densities, no memory effect, and a slow loss of charge when not in use.

Beyond consumer electronics, lithium-ion batteries are growing in popularity for electric vehicles. Research is yielding a stream of improvements to traditional LIB technology, focusing on energy density, durability, cost, and intrinsic safety.

LITHIUM IRON PHOSPHATE batteries (LFP) have one key advantage over other lithium-ion batteries is the superior thermal and chemical stability, which provides better safety characteristics than lithium-ion batteries with other cathode materials. Due to significantly stronger bonds between the oxygen atoms in the phosphate (compared to the cobalt), oxygen is not readily released, and as a result, lithium iron phosphate cells are virtually incombustible in the event of mishandling during charge or discharge, and can handle high temperatures without decomposing. LFP batteries are used by electric vehicles manufacturer Smith Electric Vehicles to power its products. Lithium Iron Phosphate chemistry also offers a longer cycle life over standard lithium ion cells. The use of phosphates also reduces the cost and environmental concerns of

Cobalt cells, particularly in regards of cobalt entering the environment through improper disposal, with considerably increased safety over the cobalt chemistry type of lithium battery.

LITHIUM–SULFUR BATTERIES (Li–S) are a rechargeable galvanic cell with a very high energy density. By virtue of the low atomic weight of lithium and moderate weight of sulfur, Li–S batteries are relatively light; about the density of water. They were demonstrated on the longest and highest-altitude solar-powered airplane flight in August, 2008. Lithium–sulfur batteries may succeed lithium-ion cells because of their higher energy density and the low cost of sulfur. There is much interest in using them for electric vehicles.

LITHIUM-TITANATE BATTERIES are a type of rechargeable battery, which have the advantage of being faster to charge than other lithium-ion batteries. Some analysts speculate that lithium-titanate batteries will power electric cars of the future.

SODIUM-ION BATTERY is a new technology developed by a Pittsburgh company uses simple chemistry—a water-based electrolyte and abundant materials such as sodium and manganese—and is expected to cost $300 for a kilowatt-hour of storage capacity, less than a third of what it would cost to use lithium-ion batteries. Tests have shown that the battery can last for over 5,000 charge-discharge cycles and has an efficiency of over 85 percent.

ELECTRIC DOUBLE-LAYER CAPACITORS are being developed to achieve the energy density of lithium ion batteries, offering almost unlimited lifespans and no environmental issues. These capacitors (batteries) could be an improvement over lithium-ion energy density several times if they can be produced. At present, the best ultracapacitors can only store about 5 percent of the energy that lithium-ion batteries hold, limiting them to a couple of miles per charge.

LIGHT DELIVERY TRUCKS

Picture at left is one of several electric trucks used by La Brea Bread for local short haul deliveries. Short haul trucks such as postal delivery and UPS show potential for electric power when they operate on a short and consistent route. Electric vehicles go back to the beginning of the 20th century when the Railway Express Agency used electric trucks on short routes within a city. Short haul trucks are presently being used in many parts of the world where the high cost of liquid fuels make electric vehicles practical.

Staples ordered 41 Smith Electric trucks to deliver office supplies. Frito-Lay ordered 176 trucks to deliver their delicious snacks. These trucks manufactured by Smith Electric use Lithium Iron Phosphate batteries (LFP)

HEAVY TRUCKS

Heavy trucks primarily use diesel derived from oil. Attempts are being made to develop biofuels for trucks, but large scale, long term prognosis does not appear to be favorable. Long haul trucks have been switching to piggy-back railway as a means of reducing their fuel costs due to increasing prices of diesel fuel. This trend will tend to increase as the cost of diesel rises. With more rail shipments, new piggy-back rail terminals will be built further encouraging the use of rail transport. More piggy-back rail terminals will require large investments which will become economically practical as the cost of fuel increases.

For use at truck terminals, Balqon Corporation produces an All Electric Terminal Tractor that can carry loads up to 60,000 lbs. It is designed to transport containers at shipping ports and large warehouses. It has a range of 95 miles on a charge and a speed of 25 miles per hour. It is powered with a Lithium Iron Phosphate battery pack that has an extended life cycle of 3000 cycles at 80 percent DOD or 5000 cycles at 70 percent DOD. Because the drive system runs quiet, operator stress and fatigue are minimized. They are 76% More Energy Efficient than comparable Diesel trucks.

CHARGING ELECTRIC VEHICLES

Gasoline-free is becoming a reality for many drivers. Cars like the Nissan Leaf can cover considerable distances, providing sufficient range for local errands and most daily commutes.

Charging the car's battery pack at home, or topping up at the office or shopping mall, will work fine for most drivers. For trips beyond the range of a single battery charge, it will be necessary to pull up to a charging kiosk and plug in for a rapid charge. There are public charging stations in service now, and new ones are coming online daily, but those typically take several hours to fully replenish a battery.

New electric vehicle batteries have the ability for quick battery boosts. Using a direct current fast charger, the Nissan Leaf can refill to 80 percent capacity in 30 minutes. Fast charging is becoming an essential design feature if the automobile manufacturers expect to sell their vehicles.

Universal fast charging methods and associated electrical connections will be worked out so that today's prevalent D.C. fast-charge systems are built to a standard developed in Japan by Nissan, Mitsubishi and Subaru in conjunction with Tokyo Electric Power. Eventually, this standard, called Chademo, may prevail as the universal standard for charging. Other groups, including the American National Standards Institute, are also working on standards and codes for electric cars.

Leisurely overnight recharging is no problem. All electric cars come with a standard charging cable that can plug into a common 120-volt household electrical outlet.

FUEL CELLS

A fuel cell is an electrochemical device that converts chemical energy from a fuel into electric energy that can be used to power an electric motor to drive a car. The main drawback of using a fuel cell in a car is the prohibitive cost of fuel cells. Electricity is generated from the reaction between a fuel supply and an oxidizing agent. The reactants flow into the cell, and the reaction products flow out of it, while the electrolyte remains within it. Fuel cells can operate continuously as long as the necessary reactant and oxidant flows are maintained.

Fuel cells are different from conventional electrochemical cell batteries in that they consume reactant from an external source, such as hydrogen or natural gas.

Many combinations of fuels and oxidants are possible. A hydrogen fuel cell uses hydrogen as its fuel and oxygen as its oxidant. Fuel cells are many times more expensive than rotary type generators to develop electricity. For that reason, fuel cells will be limited to applications where small size and reliability are the primary necessity.

NATURAL GAS TO ELECTRICITY FUEL CELL

The Natural Gas Fuel Cell (NGFC) is an emerging technology that uses an electrochemical process to generate electrical power using natural gas, or other gas such as hydrogen as the fuel.

Fuel cells can provide high quality power for a variety of uses. The fuel cell technology can save energy when generating electricity due to the high efficiency of the electrochemical process and by using low cost natural gas. Natural gas fuel cells are starting to be used as back up when electric utilities cannot meet demand re-quirements.

The negative side of natural gas fuel cells is that natural gas can be used directly in reciprocating engines, and natural gas is second in line to be depleted after oil, leaving expensive hydrogen as the input for fuel cells. Therefore;

If future technology provides us with sufficient renewable electricity from wind power, nuclear energy and geothermal sources, Liquefied Hydrogen may evolve as one of the mobility fuels of the future.

HYDROGEN TO ELECTRICITY FUEL CELL

Similar to natural gas fuel cells, hydrogen fuel cells generate electricity which can be used to power vehicles. They are more efficient than internal combustion engines, but expensive to build. Small fuel cells can power electric cars. Large fuel cells can provide electricity in remote places with no power lines.

Because of the high cost to build fuel cells, large hydrogen power plants won't be built for a while. However, fuel cells are being used in some places as a source of emergency power, from hospitals to wilderness locations. Portable fuel cells are being sold to provide longer power for laptop computers, cell phones, and military applications.

Today, there are more than 300 hydrogen-fueled vehicles in the United States. Most of these vehicles are buses and automobiles powered by electric motors. They store hydrogen gas or liquid on board and convert the hydrogen into electricity for the motor using a fuel cell.

Present cost of fuel cell vehicles greatly exceeds that of conventional vehicles in large part due to the expense of producing fuel cells. Hydrogen vehicles are starting to move from the laboratory to the road. Hydrogen vehicles are in use by a few state agencies and a few private entities.

As of 2010 there were only 68 hydrogen refueling stations in the United States, nearly a third of which were located in California. There are so-called "chicken and egg" questions that hydrogen developers are working hard to solve, including: who will buy hydrogen cars if there are no refueling stations? And who will pay to build a refueling station if there are no cars and customers? Hydrogen has been called one of the least efficient and most expensive possible replacements for gasoline in terms of reducing greenhouse gases with other technologies that are less expensive and more quickly implemented. A comprehensive study of hydrogen to power transportation has found that there are major hurdles on the path to achieving the vision of the hydrogen economy; the path will not be simple or straightforward. The Ford Motor Company has dropped its plans to develop hydrogen cars, stating that "The next major step in Ford's plan is to increase over time the volume of electrified vehicles".

COMPRESSED AIR

Tata Motors, a division of India's largest conglomerate, has announced it will start selling a car in India that stores energy in the form of compressed air, as opposed to gasoline, diesel or electric batteries. Not the usual run-of-the-mill car we've known for over 100 years. No. This one runs on air. Yes. The type we breathe in and out every day, the one found almost everywhere on earth.

The five-seater OneCAT (irony) should weigh about 350kg. Work has already started, according to the inventor of the engine system to be used, a Frenchman by the name of Guy Negre. He has been working on this concept for over ten years, and says it's finally ready for production. Tata is just the right partner for him to go with. Tata is the only licensee allowed to sell the car, and for now sales will be limited to India.

The technology has over 30 patents registered. Driving the wheels will be a sort of engine powered by compressed air which is stored in carbon fibre tanks. A top speed of 70 mph can be reached, but is probably not advisable.

Longer trips are backed up by a fuel burner that heats up the air to better boost the pistons. This fuel burner will use all kinds of liquid fuel, according to Negre's company. Fuel economy of 120 mpg on these trips is attainable, while town driving, because it will use air, should be even cheaper.

Negre explains that safety is paramount on such a project, that during a crash the air tanks will not shatter but will split with a very loud bang, with the biggest risk there being to the ears.

Compressed air cars use electrical energy to compress the air that is used for propulsion. When that electrical energy is obtained from renewable sources such as hydroelectric or wind power, the compressed air car is considered environmental friendly. Another advantage is that air can be compressed at off-peak hours and stored in large tanks ready for use the next day.

Refueling the compressed air container directly from a home or low-end conventional air compressor may take as long as 4 hours, though specialized equipment at service stations may fill the tanks in only 3 minutes.

A US company is developing a compressed air car. Magnetic Air Cars Inc. 521 Charcot Avenue, San Jose, CA claims that they are building an electric car with compressed air tanks instead of expensive rechargeable batteries that can only provide short driving distances. They believe their engineers have developed a solution for manufacturing air while the vehicle is in motion which may provide an unlimited driving range. They use only one specially designed battery the size of a regular car battery which has three times the charging capacity and a much longer life cycle. Instead of $3,000-$5,000 for lithium batteries, they claim their battery costs less than $70. Wow.

HYDRAULIC ACCUMULATORS

UPS in partnership with the US Environmental Protection Agency, will test a small fleet of hydraulic hybrid delivery trucks in the United States. The new vehicles can achieve 50-70% better fuel economy, a 40% reduction in greenhouse gas emissions, and pay for their extra expense in less than 3 years.

UPS will field two hydraulic hybrids in Minneapolis, MN, in early 2009 and an additional five hydraulic hybrid trucks will be deployed later in 2009 and early 2010. Although this sounds like a tiny fleet, keep in mind that this is the largest scale commercial test of hydraulic hybrids ever conducted.

The UPS hybrid hydraulic truck is a standard-looking 24,000 pound package, with an EPA-patented diesel series hydraulic hybrid drive attached to the rear axle. In a series hydraulic hybrid, the conventional drivetrain is replaced with a hydraulic system that stores energy by compressing gas in a chamber using hydraulic fluid. It works in much the same way that a hybrid electric car does — a small, efficient motor generates power which gets stored for later use — only, the way energy is stored in a hydraulic hybrid is in a pressurized chamber rather than in a battery.

Sounds like another version of compressed air storage.

The hydraulic hybrid drivetrain eliminates the need for a conventional transmission and increases fuel economy in three ways: A large amount of the energy that is otherwise wasted in braking can be recovered to pressurize the hydraulic fluid.
The engine operates much more efficiently — similar to a hybrid electric car, only without the bulky batteries
The engine can easily be shut off and instantaneously restarted during regular driving — such as when the vehicle is slowing down or stopped at a light – or making a delivery.

UPS has been developing what it calls its "green fleet" over the last several years and currently has more than 1,600 low carbon emissions vehicles including electric, hybrid-electric, compressed natural gas, liquefied natural gas, and propane trucks.

PASSENGER TRANSIT

ELECTRIC BUSSES began in the summer of 2000, to serve the Hong Kong Airport, operating a 16-passenger Mitsubishi Rosa electric shuttle bus, and in the fall of 2000, New York City began testing a 66-passenger battery-powered school bus, an all-electric version of the Blue

Bird TC/2000. A similar bus was operated in Napa Valley, California for 14 months ending in April, 2004.

The 2008 Beijing Olympics used a fleet of 50 electric buses, which have a range of 130 km (81 mi) with the air conditioning on. They use Lithium-ion batteries, and consume about 1 kW·h/mi. The buses were designed by the Beijing Institute of Technology and built by the Jinghua Coach Co. Ltd. The batteries are replaced with fully charged ones at the recharging station to allow 24 hour operation of the buses.

Seoul Metropolitan Government runs the world's first commercial all-electric bus service. The bus was developed by Hyundai Heavy Industries and Hankuk Fiber which make body as light as possible from carbon composite material. Provided by Li-on battery and Regenerative Braking Systems, the bus may run to 52 miles in a single 30 minutes charge. The maximum speed is 62 miles/hour.

THE TROLLEYBUS is a type of electric bus powered by two overhead electric wires, with electricity being drawn from one wire and returned via the other wire, using two roof-mounted trolley poles.

A trolleybus is an electric bus that draws its electricity from overhead wires (generally suspended from roadside posts) using spring-loaded trolley poles. Two wires and poles are required to complete the electrical circuit. This differs from a tram or streetcar, which normally uses the track as the return part of the electrical path and therefore needs only one wire and one pole (or pantograph). They also are distinct from other kinds of electric buses, which usually rely on batteries.

Currently, around 315 trolleybus systems are in operation, in cities and towns in 45 countries. Altogether, more than 800 trolleybus systems have existed, but not more than about 405 concurrently.

CAPABUS is an experimental new form of electric bus, which runs without continuous overhead lines by using power stored in large onboard electric double-layer capacitors, which are quickly recharged whenever the vehicle stops at any bus stop to get fully

charged. The buses have very predictable routes and need to stop regularly, every 3 miles (4.8 km), allowing opportunities for quick recharging. The trick is to turn some bus stops along the route into charging stations. At these stations, a collector on the top of the bus rises a few feet and touches an overhead charging line. Within a couple of minutes, the ultracapacitor banks stored under the bus seats are fully charged.

This technology could be used to electrify trucking routes along major highways, but instead of stopping to recharge, trucks would continuously recharge using overhead trolley wires. They would run on the double-layer capacitors on sections of the highway where it is not be feasible to have overhead trolley wires such as at intersections, under existing low bridges and truck service stops.

THE GAPBUS is a bus without rails or surface power lines. Power is supplied over a gap of up to 4.7 inches from a power line embedded in the ground. This principal of induction to transfer electrical energy has been used in many forms including mono-rail trains and can be adapted to highways so that vehicles other than busses and trains can be powered by electrical energy.

Induction transfer of electrical energy is technically possible, but it will have to compete with overhead power lines for future electrification of our transportation systems.

PASSENGER RAIL

THE NEW YORK CITY RAIL TRANSIT SYSTEM is the largest in the world. By far the dominant mode of transportation in New York City is rail. Only 6% of shopping trips in Manhattan's Central Business District involve the use of a car. The city's public transportation network is the most extensive and among the oldest in North

America. The New York City Subway is the largest rapid transit system in the world when measured by track mileage (656 miles of mainline track), and the fourth-largest when measured by annual ridership (1.4 billion passenger trips in 2005). It is the second-oldest subway system in the United States after the rapid transit system in Boston. In 2002, an average 4.8 million passengers used the subway each weekday. Life in New York City is so dependent on the subway that the city is home to two of only four 24-hour subway systems in the world

The New York City Subway system runs entirely on electricity because most of the trains operate underground where pollution from engine exhausts cannot be tolerated.

AMTRAK NORTHEAST CORRIDOR between Boston - New York - Philadelphia - Washington, DC provides rail service at speeds up to 150 mph. High-speed rail in the United States is extremely limited. Although the United States has large areas with population densities comparable to Western Europe and East Asia, there exists only one high-speed rail line. Moreover, this line is considerably slower than high-speed rail lines provided in most other developed societies. In 2000 Amtrak introduced the *Acela Express*, which operates at a maximum speed of 160 mph between Washington, D.C. and Boston. These trains tilt into curves along the track, allowing them to travel between Washington and New York in 2 hours and 45 minutes. This time—an average speed of only 83 mph--is heavily influenced by the need to travel through and around Baltimore on tracks which are in some places over 160 years old. Ambitious long-term plans by the Maryland DOT to create a new express route through Baltimore for Acela and MARC commuter trains would significantly reduce this travel time.

Unlike Asian or European systems, the Acela shares its tracks with conventional rail, and thus is limited to an average speed of 68 mph for the entire distance with brief segments up to 150 mph. A federal allocation of $8 billion for high-speed rail projects as a part of the 2009 stimulus package has prompted U.S. federal and state planners to coordinate the expansion of high-speed service to ten other major rail corridors.

Passenger rail such as the Amtrak Northeast Corridor powered by electricity (obtained from nuclear and geothermal sources) may be the only alternative as fossil fuels become deleted.

Amtrak's nationwide network runs mostly on tracks owned by freight rail companies like Union Pacific, BNSF, and CSX. While BNSF does a pretty good job of expediting Amtrak passenger trains, Union Pacific, in particular, seems to do its utmost to delay and interfere with Amtrak trains; hence some of the nicknames like Unlimited Parking and Utterly Pathetic. Adding ANY additional Amtrak trains on freight railroad tracks would require agreement by those freight railroads.

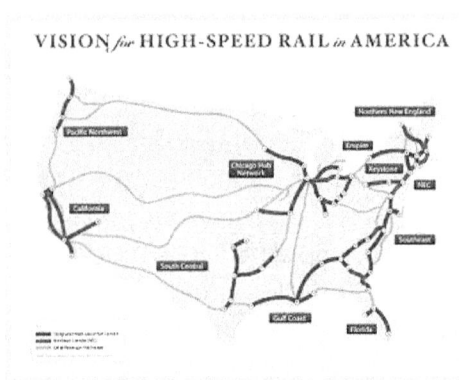

VISION for HIGH-SPEED RAIL in AMERICA

HIGH SPEED RAIL plans for the United States were announced by President Obama on April 16, 2009.

The Administration has placed new emphasis on building high-speed and intercity passenger rail to connect communities and economic centers across the country. A fully de-veloped passenger rail system will complement highway, aviation and public transit systems.

With the successful completion of the original phases of the Northeast Corridor (NEC) Improvement Project offering Amtrak's maximum 150 mph Acela train service between Washington, New York, and Boston, efforts to develop high-speed intercity passenger rail service have expanded beyond the NEC. However, just as the Interstate Highway System took 50 years to complete, the true potential of a fully integrated high-speed intercity passenger rail network will not be achieved or realized overnight.

The Department of Transportation is working with States to plan and develop high-speed and intercity passenger rail corridors that range from upgrades to existing services to entirely new rail lines exclusively devoted to 150 to 220 mph trains. Implementing these corridor projects and programs will serve as a catalyst to promote economic expansion (including new manufacturing jobs), create new choices for travelers in addition to flying or driving, reduce national dependence on oil, and foster livable urban and rural communities.

HIGH-SPEED RAIL (HSR) is a type of passenger rail transport that operates significantly faster than the normal speed of rail traffic. Specific definitions by the European Union include 124 mph for upgraded track and 155 mph or faster for new track, while in the United States, the U.S. Department of Transportation defines it as "reasonably expected to reach sustained speeds of more than 125 mph", although the Federal Railroad Administration uses a definition of above 110 mph.

This photograph shows the Shinkansen between Tokyo and Osaka in Japan which in 1993 reached a speed of 264 mph.

European and Asian commuter railroads are far ahead of the United States in electrification and speed. Electrification of railroads in the United States will take large investments and years to accomplish. When railroad executives realize what the future cost of fossil fuels will be, they may start to plan electrification of their rails.

The following table shows all high speed dedicated lines (speed over 155 mph) in service and under construction as of 2010.

Country	In Operation (km)	Under Con- struction km)	Total Country (km)
China	6,158	14,160	20,318
Spain	2,665	1,781	3,744
Japan	2,118	377	2,495
France	1,872	730	2,602
Germany	1,032	378	1,410
Italy	923	92	1,015
Russia	780	400	1,180
Turkey	457	591	1,048
Taiwan	345	0	345
South Korea	330	82	412
Belgium	209	0	209
Netherlands	120	0	120
United Kingdom	113	0	113
Switzerland	35	72	107

Note that the United States does not have any rail lines that meet the criteria of a high speed dedicated line.

Note that China has more high speed rail than the rest of the world combined.

Unfortunately, due to the accident that took place when two bullet trains collided in Zhejiang Province, China on July 23, 2011, killing about 40 people, China suspended all new high-speed railway projects, indefinitely.

COMMERCIAL RAIL

RAIL SHIPMENT of goods is the second most cost effective way to ship freight long distance. The most cost effective method of shipment is by water. Fastest growing rail traffic segment is currently intermodal which is the movement of shipping containers or truck trailers by rail and at least one other mode of transportation, usually trucks or ocean-going vessels. Intermodal combines the door-to-door convenience of trucks with the long-haul economy of railroads. Rail intermodal has tripled in the last 25 years. It plays a critical role in making logistics far more efficient for retailers and others.

Today, most rail carriers can move a ton of freight nearly 500 miles on average, per gallon of diesel fuel. Compared to this, a semi-trailer will carry on average, a ton of freight 80 miles on a gallon of diesel fuel.

The efficiency of intermodal provides the U.S. with a huge competitive advantage in the global economy. Freight railroads offer major public benefits in addition to cost-competitiveness and efficiency. First, railroads are more fuel efficient than other modes of transportation. On average, they are three times more fuel efficient than trucks. In 2006, railroads moved a ton of freight an average of 436 miles per gallon of fuel. That number is an 80 percent increase from 1980. Because of their fuel efficiency, railroads also have a clear advantage over other modes of transportation in terms of greenhouse gas emissions, most notably carbon dioxide.

Highway congestion costs $78 billion per year just in wasted travel time (4.2 billion hours) and wasted fuel (2.9 billion US gallons). Railroads may change traffic congestion by replacing some trucks currently transporting goods on our highways.

PASSENGER SHIPS

The period between the end of the 19th century and World War II was known as the "golden age" of ocean liners. With the advent of high speed jet aircraft, there was no longer a need for ocean liners as a means of transportation. Ocean travel gave way to cruise ships that maintained the romance of luxury travel on the high seas. Technology gave us larger ships with ever greater luxury. The photograph on the left shows the Oasis of the Seas which carries 5,400 passengers plus crew.

Competition between cruise ships lines resulted in cruising prices that could be afforded by mass markets. The number of ships built for cruising can only be described as enormous. The problem of the high cost of fuel consumption is starting to show its ugly head. One method of conserving fuel has been demonstrated by commercial shipping is to slow down.

Tabulation below shows the fuel consumption of typical cruise ships in nautical miles per gallon of fuel, for comparison with other forms of travel. Tests have proven that fuel consumption, pollution and costs can be improved greatly by lowering the speed from about 20 knots to 10 knots. However, from a practical point, lowering the cruising speed would give the passengers fewer ports to visit resulting in less value for their dollar. The only way this could be implemented is to have agreement between all competing cruise lines that they would all lower their cruising speed. Until ships are powered by nuclear reactors, slower speeds will be necessary.

	Passengers	Feet per Gallon of Fuel	Nautical Miles per Passenger per Gallon
Queen Elizabeth 2	1,800	49	14.5
Zandaam	1,440	73	16.8
Oasis of the Seas	5,400	18	16.0
Disney Magic	1,750	57	16.4

SOURCE – CYBERCRUISES.COM

PASSENGER PLANES

Commercial aircraft have seen a steady increase in load factor, and efficiency from 1960 through 2010. The table below shows averages for all commercial airlines.

Why do aircraft get better fuel economy than a ship? Thin air, particularly at high altitudes of 30,000 to 40,000 feet has relatively little air resistance compared to a ship that must push through tons of water.

	1960 Nautical Miles per Passenger per Gallon	2010 Nautical Miles per Passenger per Gallon
Domestic Seat	26.7	60.2
International Seat	23.6	52.1
Domestic Load Factor	58.5 %	81.9 %
International Load Factor	62.2 %	80.7 %

SOURCE RITA TABLE 4-21

Commercial aircraft fuel consumption as of 2010 is shown in the table below. Compare these figures to the fuel consumption of a cruise ship.

Aircraft	Passengers	Range Nautical Miles	Fuel Used Gallons	Nautical Miles per Passenger per Gallon
Airbus 321-200	185	3,000	6,350	87
Airbus 340-600	419	7,750	51,750	63
Boeing 737-400	146	4,200	6,300	97
Boeing 747-400	524	7,260	57,200	66
Boeing 767-300	270	4,260	16,700	69
Boeing 787-9	290	8,200	36,640	65

SOURCE - AIRLINES.NET

Nautical miles per passenger per gallon is generally greater for short range airplanes such as the Airbus 321 and the Boeing 737 because they can take off with less fuel for the shorter distance.

Fuel is the number one expense for the airline industry accounting for up to 50 percent of and airline's operating costs. The good news is that airlines are making great progress to conserve fuel. The bad news is that airlines are dependent upon jet fuel with high energy content per pound (specific energy). When we reach oil depletion, replacement fuel for airplanes such as Ethanol and Hydrogen will be expensive and may not have the energy content necessary to carry full passenger loads for long distances. Algae fuel has been tested on both aircraft and ships and passed all performance tests. Both the airline and shipping industries are watching for the price of Algae Fuel to become competitive with fuels derived from oil. When this occurs, the entire outlook on oil depletion and renewable energy may change.

COMMERCIAL SHIPS

As of 2010, the world ocean-going fleet was made up of 50,000 ships. Number of ships by sector:

General Cargo Ships	16,224
Bulk Carriers	8,687
Container ships	4,831
Tankers	13,175
Passenger ships	6,597

The world ocean going fleet consumes a total of 7.723 million barrels of oil per day. Based upon total world oil consumption of 82.769 million barrels per day, this accounts for 9.4% of total world oil consumption.

A combination of the 2008 recession and growing awareness in the shipping industry about climate change emissions encouraged many ship owners to adopt "slow steaming" to save fuel starting about 2009. Lowering of speeds from the standard 25 knots to 20 knots has resulted in enormous fuel savings, but many major companies have now taken this a stage further by adopting "super-slow steaming" at speeds of 12 knots to further conserve fuel.

Photo to the left is the Emma Maersk Triple E class container ship, the largest ever built. It is 1,300 feet long capable of carrying 18,000 twenty foot equivalent container units (TEU). As of 2011, there were a total of 20 of these ships on order adding to the Maersk fleet of 1,300 ships. Future Maersk Triple E Class, as well as many large tankers (ULCCs) and some bulk carriers (VLOCs) will not be able to pass through the new, much larger locks because they are too wide for even the new locks. These ships are 194 ft wide while the new locks will be only 180 ft wide.

Photo to the left is a Batillus class super tanker with a length of 1,359 ft and width of 207 ft. Consistent with the limiting dimensions of the new Panamax locks, these super tankers are being phased out in favor of ships that can pass through the new Panamax locks. In 2005, 2.42 billion metric tons of oil were shipped by tanker. This amounted to 34.1% of all seaborne trade for the year. In 1970 1.44 billion metric tons of oil were shipped by tanker. This accounted to an increase of 68% in 7 years. This increase is typical all of ocean shipping for that period.

The Panama Canal plays a key factor in shipping around the world with size limitations referred to as Panamax dimensions which are determined principally by the dimensions of the canal's lock chambers, 110 ft wide by 1,050 ft long, and 41.2 ft deep.

The future Panamax locks will have dimensions of 180 ft wide by 1,400 ft long, and 60 ft deep. Completion of the new Panamax locks is expected in 2014.

NUCLEAR NAVY

The United States has the strongest military force in the world. In addition to training the best fighting men, the military has the responsibility and the funding to develop the best equipment. Going back to World War II, the military developed the best aircraft and has stayed ahead of the rest of the world ever since. The navy saw the need for nuclear ships so they developed the largest fleet of nuclear powered naval vessels.

Admiral Hyman G. Rickover, in the 1940s was the driving force for the development of nuclear naval ships. The first nuclear-powered submarine, USS Nautilus, put to sea in 1955. Other submarines followed, powered by single reactors, and a cruiser, USS Long Beach, in 1961, powered by two reactors. The aircraft carrier USS Enterprise, commissioned in 1961, is powered by eight reactor units.

By 1990 there were more nuclear reactors powering ships (mostly naval) than there were nuclear reactors generating electric power in commercial power plants worldwide.

Russia has built four nuclear cruisers and nine ice breakers. Russian nuclear ice breakers are used to force through the ice for the benefit of cargo ships and other vessels along the northern seaway. In all, ten civilian nuclear powered vessels have been built in Russia. Nine of these are icebreakers, and one is the container ship Sevmorput with an ice-breaking bow.

NUCLEAR COMMERCIAL SHIPS

As of 2010, commercial ships world-wide consumed 7.723 million barrels of oil per day, accounting for 9.4% of total world oil consumption. Because of this, consideration has been given to the possibility of nuclear power for ships. The United States, Germany, Japan and Russia have each built a nuclear powered commercial ship, all with negative results.

The negative results can be correlated with the three major nuclear accidents starting with Three Mile Island, Chernobyl, and then Fukushima.

NS Savannah, the United States first and only nuclear powered merchant ship was designed in hopes of finding peaceful uses for Nuclear energy as part of the Atoms for Peace program. During her short 5 years of service from 1965 to 1970, she sailed over 450,000 miles saving over 29 million gallons of fuel oil but her high maintenance cost led to her downfall. The program was conducted prior to March 1979, the time of the Three Mile Island meltdown, and therefore, nuclear safety was not the primary cause of the termination of the NS Savannah.

NS Otto Hahn, the German-built nuclear powered cargo vessel was powered by a single 38 MW reactor which first achieved criticality in 1968. The Otto Hahn made its first port call in Casablanca in 1970 and continued to operate under nuclear power until 1979. In nine years, the Otto Hahn steamed 650,000 nautical miles and visited 33 ports in 22 countries. This program also was conducted prior to March 1979, and therefore, nuclear safety was not the primary cause of the termination of the program.

Mutsu was Japan's first and only nuclear-powered ship. It was built as a nuclear merchant ship, but never carried commercial cargo. On September 1, 1974, as the crew brought the reactor up to 1.5% of capacity there was a minor leak of neutrons and gamma rays. Between 1978 and 1982, various modifications were made to the reactor shield of the Mutsu. She then completed her original design objective of travelling 82,000 kilometres (51,000 mi) in testing, and was decommissioned in 1992. Termination of this program can be contributed to both economic reasons and safety concerns since it occurred after March 1979, when Three Mile Island had a meltdown.

Russia built a nuclear powered commercial ship, the Sevmorput. She was launched on 20 February 1986. After leaving the shipyard the Sevmorput, she sailed without nuclear power through the Mediterranean and around Africa. Authorities in several ports refused to accept the two-month-old ship into their ports due to popular protests. In addition the harbor workers refused to load or unload any cargo or provide any port services due to fears of radiation leakage. This was caused by uncertainty about the safety of the ship's nuclear propulsion system in the shadow of the Chernobyl disaster only few years earlier. The Sevmorput was still in service in 2011, but was never converted to nuclear power

Russia is well advanced with plans to build a floating nuclear power plant for their far eastern territories. The design has two 35 MWe units based on the KLT-40 reactor used in icebreakers (with refueling every four years). Some Russian naval vessels have been used to supply electricity for domestic and industrial use in remote far eastern and Siberian towns.

TRANSPORTATION ALTERNATIVE FUEL SOURCES

GASOLINE FROM COAL

Coal conversion to synthetic fuels was developed in Germany in 1923 known as the Fisher – Tropsch process. During World War II, Germany used substitute oil products by the Bergius process, the Fischer–Tropsch process and other methods.

As of 2009 worldwide commercial synthetic fuels production capacity was over 240,000 barrels per day with numerous new projects in construction or development.

In the United States, the Adams Fork Energy project will convert regional coal into premium grade ultraclean gasoline, producing 18,000 barrels per day. When fully developed the Adams Fork project will be the largest coal to gasoline project in the world.

During the operation of this facility, air emissions are expected to be so low that it will qualify as a minor source under US law. On the environmental side, the polluting properties of coal, starting with mining and lasting long after burning, and the large amounts of energy required to liquefy it means that liquid coal produces more than twice the global warming emissions as regular gasoline and almost double that of ordinary diesel. As pundits have pointed out, driving a Prius on liquid coal makes it as dirty as a Hummer on regular gasoline.

One ton of coal produces only two barrels of fuel. In addition to the carbon dioxide emitted while using the fuel, the production process creates almost a ton of carbon dioxide for every barrel of liquid fuel. Which is to say, one ton of coal in, more than two tons of carbon dioxide out. Proponents of coal-to-liquid plants argue that the same technologies that may someday capture and store emissions from coal-fired plants will also be available to coal-to-liquid plants. But even if the carbon released during production were somehow captured and sequestered, a technology that remains unproven at any meaningful scale, liquid coal would still release 4 to 8 percent more global warming pollution than regular gasoline.

GASOLINE FROM NATURAL GAS

A Texas company claims that it has developed a cheaper and cleaner way to convert natural gas into gasoline. The company behind the technology, Dallas-based Synfuels International, says that the process uses fewer steps and is far more efficient than more established techniques based on the Fischer-Tropsch process.

Synfuels claims their refinery could produce a barrel of gasoline for $25. Under current fuel prices, such a plant could pay for itself in as little as four years.

Texas A&M University licensed its approach to Synfuels and partly owns the company, which has been operating a $50 million demonstration plant in Texas since 2005 and says that it is close to signing a deal for its first commercial refinery near Kuwait City. But Synfuels isn't alone in trying to make GTL more economical.

Producing gasoline from coal and natural gas will accelerate depletion of our coal and natural gas resources. The long term solution to oil depletion will be renewable bio-fuels such as Ethanol and Algae Fuel.

ETHANOL

ETHANOL (ethyl alcohol) is produced by fermenting renewable crops like corn or sugarcane in a process similar to making whiskey, vodka and gin. The process requires large amounts of arid land and electricity. Opponents claim that Ethanol production consumes six BTU of electrical energy (renewable) to produce just one BTU contained in Ethanol. This may be true to a large extent, but we have very little alternative at the present time. Ethanol is a transportation fuel, which is a fuel that can be conveniently carried in a vehicle. Opponents of Ethanol claim that there are other transportation fuels that can be made from natural gas and coal. True again, but natural gas and coal will become depleted soon after oil becomes depleted. The word "soon" can mean 50 years or it can mean 500 years. If we start making transportation fuels from natural gas and coal, the word "soon" will become closer to 50 years than to 500 years.

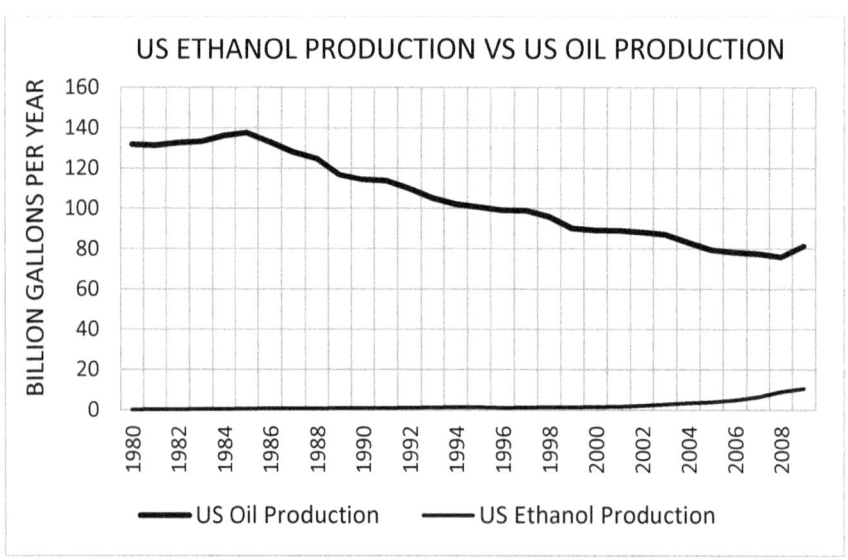

US ENERGY INFORMATION ADMINISTRATION
INDEPENDENT STATISTICS & ANALYSIS TABLE 5.1
US ETHANOL PRODUCTION FROM BIOFUELS JOURNAL

Ethanol is the most commonly used biofuel, generally used as an additive in gasoline. In the above chart, if we assume that U. S. oil production and US ethanol production will continue the present trend, and we extrapolate to the year 2030, U. S. ethanol produc-

tion will exceed US oil production. Although present trends probably will not continue indefinitely, the chart does give an indication
of the rapid increase in the use of ethanol in the United States,
Brazil and the World.

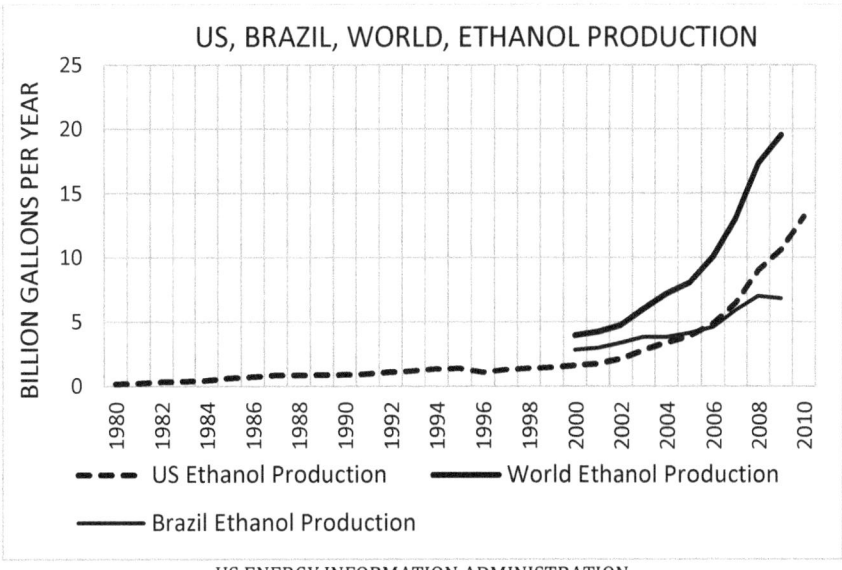

US ENERGY INFORMATION ADMINISTRATION
INDEPENDENT STATISTICS & ANALYSIS TABLE 5.1
US ETHANOL PRODUCTION FROM BIOFUELS JOURNAL

In the above chart, statistics for World ethanol production and
Brazil ethanol production prior to 2000 are not readily available.
The above chart also shows that about 2006, the United States
surpassed Brazil as the largest producer of ethanol. The U.S. produced 13.2 billion US gallons of ethanol fuel in 2010.

Controversy continues over the use of ethanol as a replacement
for gasoline. Since most U.S. ethanol is produced from corn and
the required electricity from many distilleries comes mainly from
coal plants, there has been considerable debate about how sustainable corn-based bio-ethanol could be in replacing fossil fuels in
vehicles. There is controversy relating to the large amount of
arable land required for crops and its impact on grain supply.

THE ETHANOL HOAX

One gallon of conventional gasoline has 5.253 million BTU, while one gallon of Ethanol has only 3.563 million BTU. When conventional gasoline is blended with 10% Ethanol, the blended gasoline will have less than 97% of the energy of conventional gasoline so that an automobile rated at 30 MPG will achieve only 29 MPG when using 10% Ethanol, a 3% reduction in MPG. Ask your car salesman if the gasoline mileage for a particular car is rated using straight gasoline or a blend of 10% Ethanol. Good luck.

BIOFUELS

BIOFUEL is a category of fuel which is in some way derived from biomass. The term covers solid biomass, liquid biofuels and various biogases. Bioenergy ranks second to hydropower in renewable U.S. primary energy production and accounts for 3 percent of the primary energy production in the United States. It is environment friendly and is a sustainable resource. Some crops are planted and grown expressly for the purpose. Biofuels currently produces 5 times more renewable energy for the United States than wind and solar energy combined.

BIOGASES

BIOGASES typically refer to gas produced by the biological breakdown of organic matter in the absence of oxygen. Biogas originates from manure, sewage, municipal waste, green waste, plant material and crops. Biogas comprises primarily methane and carbon dioxide. The gases methane, hydrogen and carbon monoxide can be combusted with oxygen. This energy release allows biogas to be used as a fuel. Landfills have pipes inserted into the waste to capture biogas which is pressurized and used to drive vehicles operating on the landfill. Biogas can be used in a gas engine to convert the energy into electricity and heat. Biogas can be compressed, much like natural gas, and used to power motor vehicles.

Landfills have pipes into the ground to capture the methane gas emitted from decaying waste. The gas is compressed and used to drive vehicles that are used to maintain the landfill.

Farmers accumulate animal waste under large plastic covers, collect the methane that comes from the waste, and compress it for use in farm vehicles.

ALGAE FUELS

Nature took 150 million years to form oil and natural gas from tiny aquatic animals. It will take man less than 400 years to deplete those resources. We could not have accomplished this without modern technology. The good side of the picture is that modern technology may provide us with a way to replace oil that took millions of years to form with algae oil that takes only a few days to form.

Technology got us into this energy depletion crisis and it is possible that technology may get us out of this crisis by growing algae for fuel.

President Obama, on February 23rd 2012 at the University of Miami said the following about algae: "We're making new investments in the development of gasoline and diesel and jet fuel that's actually made from a plant-like substance — algae. You've got a bunch of algae out here, right? (Laughter.) If we can figure out how to make energy out of that, we'll be doing all right. Believe it or not, we could replace up to 17 percent of the oil we import for transportation with this fuel that we can grow right here in the United States. And that means greater energy security. That means lower costs. It means more jobs. It means a stronger economy."

Algae fuels may be an alternative to fossil fuels when used with sustainable feedstock such as sugar. Several companies and government agencies are funding efforts to reduce production and operating costs to make algae fuel production commercially viable. The results of these efforts are very promising.

Rising jet fuel prices are putting pressure on airline companies, creating an incentive for algae jet fuel research. Trials have been carried out with aviation biofuel by Air New Zealand, and Virgin Airlines. February 2010, the Defense Advanced Research Projects Agency announced that the U.S. military was about to begin large-scale production of oil from algal ponds into jet fuel. November

2011, the first commercial flight using algae biofuel flew from Houston, Texas to Chicago O'Hare airport. Lufthansa Airline has signed an agreement with Algae, Tec Ltd to evaluate algae oil.

Maersk, the largest shipping company in the world understands the eventual depletion of oil that is necessary to propel its ships around the world. In a test of algae fuel, Maersk in conjunction with the U.S. Navy has agreed to sail a 300-foot container ship 6,500 nautical miles from Germany to India on about 30 tons of biofuel. The ship is already outfitted with a separate, dedicated engine, fuel tanks and blenders for Maersk's own biofuel initiative, during the test, the crew will monitor the ship's emissions and fuel efficiency.

On November 16, 2011, the United States Navy launched its largest alternative fuel test to date, pumping 20,000 gallons of Solazyme's algae-based fuel into a U.S. destroyer that was to embark on a 20-hour trip along the California coast. The USS Paul H. Foster's trip from San Diego to Port Hueneme is a part of the Navy's plan to unveil next year a small carrier strike group of small ships, destroyers, cruisers, aircraft, submarines and a carrier run on alternative fuels, including nuclear power. Naval officials will use data from an earlier trip of the same route to compare how the ship performs when running on Solazyme's biofuel.

An example of rapidly changing renewable energy technology is shown in a report made by Boeing dated January 2006, titled "Alternate Fuelled Aircraft". It discusses liquid hydrogen and ethanol as alternate fuels for aircraft with no mention of algae fuel. Less than 6 years after that report was made, aircraft were being test flown with algae fuel, not hydrogen, not ethanol.

High oil prices, competing demands between foods and other biofuel sources, and the world food crisis, have ignited interest in alga culture (farming algae) for making vegetable oil, biodiesel, bioethanol, and other biofuels, using land that is not suitable for agriculture. Among algal fuels' attractive characteristics: they do not affect fresh water resources, can be produced using ocean and wastewater, and are biodegradable and relatively harmless to the environment if spilled.

Energy from algae is attracting attention for renewable energy, with 15 small startup companies developing various systems. Energy from Algae is attracting environmentalists because when algae fuel is formed, it consumes carbon dioxide and releases oxygen. Algae fuel has several advantages over Ethanol fuel, because it takes less agricultural land to produce, requires less energy to produce and resolves the food for fuel controversy associated with Ethanol fuel.

Producing algae fuel is a complex process requiring several steps and with many variables.

    Selection of the most appropriate strain of algae
    Growing algae by one of three methods
    Harvesting algae mechanically or chemically
    Extracting oil physically or chemically
    Refining algae oil into fuel

SELECTION OF THE OPTIMUM ALGAE from over 10,000 strains including 500 culture collections registered with the World Federation for Culture Collections. Many other cultures have been developed privately and are proprietary.

There are three methods of growing algae for fuel, each with its advantages and disadvantages. It is still too early in the game to predict which method will eventually prevail.

NATURAL AND MAN-MADE PONDS are an economical method of growing algae. For millions of years, Algae has been growing in stagnant natural ponds without the intervention of man. Such ponds were considered to be contaminated and ugly. Today, Algae growing in stagnant ponds is beautiful to those producing algae fuel. The advantage of open ponds to grow algae is that it requires minimum capital equipment. The disadvantage is quality control.

Seambiotic, an Israeli firm, uses an open pond constructed in the shape of a raceway with paddle-wheels propelling the algae along the raceway providing a continuous process rather than a batch process. Cultivation growth is fed by CO_2 from a nearby power plant.

CLOSED SYSTEM algae photo bioreactor is a bioreactor that incorporates some type of light source to provide photonic energy input into the reactor. Also, an open pond could be seen as photo bioreactor, but mostly the term photo bioreactor only refers to closed systems, systems closed to the environment having no direct exchange of gases and contaminants with the environment. Specifically, algae bioreactors can be used to produce fuels such as biodiesel and bioethanol. Fundamentally, this kind of bioreactor is based on the photosynthetic reaction which is performed by the chlorophyll containing algae itself using dissolved carbon dioxide and sunlight energy.

CLOSED SYSTEM utilizing indirect photosynthesis developed by Solazyme Inc. grows microalgae in the dark, inside huge stainless-steel containers. Most microalgae produce their own nutrients by using sunlight in a photosynthetic process. Their proprietary microalgae are heterotrophic, meaning they grow in the dark (in fermenters) by consuming sugars derived from plants that have already harnessed the sun's energy. By using standard fermentation equipment, they are able to efficiently scale and accelerate microalgae's natural oil production time to just a few days and at commercial levels. They feed sugar to the algae, which the organisms then convert into various types of oil. The oil can be extracted and further processed to make a range of fuels, including diesel and jet fuel, as well as other products.

By feeding plant sugars to oil-producing microalgae in dark fermentation tanks, the technology utilizes, in contrast to common open-pond and photo bioreactor approaches. The platform is feedstock flexible and can utilize a wide variety of renewable plant-based sugars, such as sugarcane-based sucrose, corn-based dextrose, and sugar from other sustainable biomass sources including cellulosic. Solazyme's technology platform allows them to also produce and sell bio products which are made from the protein, fiber and other compounds produced by microalgae.

April 3, 2012, Solozyme Inc. and the farm-products company Bunge Ltd. formed a joint venture to produce sustainable tailored triglyceride oil for use in oleochemical and fuel applications in the Brazil market.

HARVESTING algae is the process of separating the algae from the media in which it was grown, which is mostly water. Harvesting can be done with mechanical filters, centrifugal separation, chemical separation, or a combination of these. Ultrasound and other harvesting methods are currently under development.

The optimum process that will prevail is dependent upon economics and the ability to operate continuously and reliability without intervention.

Harvesting the oil from algae-filled water is expensive, but researchers have come up with an effervescent solution where bubbles smaller than the width of a human hair can help reduce the costs of collecting algae oil. So-called micro-bubbles are already used for water purification—they surround contaminants and float them out of the liquid. Similarly, in water containing algae, bubbles can float the algae to the surface for easy collection and processing.

OIL EXTRACTION can be done physically, chemically, or a combination of both. The simplest method is mechanical crushing. When algae is dried it retains its oil content, which then can be "pressed" out with an oil press similar to extracting apple cider from apples. Mechanical crushing can be used in conjunction with chemical solvents. Osmotic shock can be used to rupture and release cellular components, such as oil. Ultrasonic extraction can greatly accelerate the extraction processes. Shock waves cause cell walls to break and release their contents into a solvent. Again, the optimum process that will prevail is dependent upon economics and the ability to operate continuously and reliability without intervention.

WASTE DISPOSAL of the algae meal, or the deoiled cake is a source of nutrients for humans and animals because the cake has high protein content, sometimes as high as 50 to 60% of dry matter. Many algal species are favorable for the nutrition of farm animals. Algae are also a rich source of carotene, vitamin C and K, and B-vitamins.

REFINING ALGAE OIL INTO FUEL utilizes existing refineries to trans esterify the purified algae oil to yield biodiesel, or hydro treat it to yield jet fuel.

Feasibility of growing algae for fuel has been demonstrated with prototype facilities and the practicality of algae fuel demonstrated by actual tests. Future prospects for algae fuel depend on production facilities that will make it cost competitive with mobility fuels derived from oil. Considering the depletion of oil which will result in increases in oil prices, the prospects for algae fuel are very promising.

Solazyme, Inc. is planning to build a pilot-scale integrated bio refinery. This will be a scale-up of their novel heterotrophic algal oil bio manufacturing process. The plant will validate the commercial scale economics of producing multiple advanced biofuels, and enable Solazyme to collect data necessary to complete the design of its first commercial scale facility.

"Fuel" is an enlightening film previously called Fields of Fuel. The 2008 documentary film won the audience award at the 2008 Sundance Film Festival. It is essential viewing for anyone interested in the future of algae fuel.

RENEWABLE ENERGY STATISTICS

Total World energy production in 2008 was 143,851 TWh, or 490,820 trillion BTU. Estimates show 78,531 trillion BTU (16%) of this total comes from renewable energy. The United States produces 7,744 trillion BTU (about 10% of the world total) of renewable energy.

Fastest growth of world renewable energy is from wind power with a growth rate of 30% annually. Next fastest growth is from Ethanol with a growth rate of 20% annually. However, if algae fuel can be produced in sufficient quantities at reasonable costs, it could replace Ethanol as the renewable mobility fuel of the future.

These rates appear to be encouraging. However, wind energy accounts for only 0.14% of total world energy production and Ethanol accounts for 0.31% of total world energy production.

The average annual increase in total world energy consumption from 1965 to 2010 was 6,670 trillion BTU per year. Increase in wind energy production from 2009 to 2010 was 1,775 trillion BTU.

These figures show that wind energy would have to increase at far greater rates than 30% annually in order to keep up with total world energy increases. Considering the practical limitations on further increases of wind energy generation and Ethanol production, these two sources of energy will not meet the future increases in energy needs of the world.

UNITED STATES RENEWABLE ENERGY PRODUCTION
TRILLION BTU

	1950	1970	1990	2009
Biomass Wood	1,562	1,429	2,216	1,891
Biomass Waste	na	2	408	447
Biofuels Ethanol	na	na	na	1,545
Hydroelectric	1,415	2,634	3,046	2,682
Geothermal	na	11	336	373
Solar and PV	na	na	60	109
Wind	na	na	29	697
Total Renewable	2,978	4,076	6,206	7,744

US ENERGY INFORMATION ADMINISTRATION
INDEPENDENT STATISTICS & ANALYSIS TABLE 10.1

HYDROELECTRIC

Hydroelectric is one of the oldest methods for generating electricity. It is economical and environmentally friendly, once land has been flooded to store the necessary water. Throughout the world, sites for future hydroelectric dams are becoming less available and less economically practical. Statistics for existing and future hydroelectric power are shown in Section Six, Energy Statistics.

WIND

Wind has been used to power pumps for reclaiming land and irrigation for many years. As of the turn of the 21st century, and the realization that fossil fuels are rapidly becoming depleted, wind energy has shown a rapid increase in development. Predictions show that by the year 2020, world electricity produced from the wind will equal or exceed that produced by hydroelectricity. Statistics for existing and future wind energy are shown in Section Six, Energy Statistics.

SOLAR

In 2007 grid-connected photovoltaic electricity was the fastest growing energy source, with installations of all photovoltaics increasing by 83% in 2009 to bring the total installed capacity to 15 GW. Photovoltaic installations in homes and businesses can take advantage of storing the electrical energy in batteries until there is a need for the electricity. Some photovoltaic installations make provision to sell the electricity to the electric grid when it is not required locally.

Germany is the world's largest consumer of photovoltaic electricity with capacity growth in 2010 in Germany accounting for 44.3% of global capacity growth. With 17.3 GW of capacity installed by the end of 2010, Germany has 43.5% of the world's solar capacity.

Despite the rapid capacity growth, share of solar power in total electricity generation world-wide remains low, at an estimated 0.1% of all renewable energy, and a small percentage of that produced by wind power.

The most negative aspect of solar energy is the production cycle which has maximum output at high noon and practically zero output after sundown. In order to make solar energy a stand-alone producer of energy there must be means of storing the energy. between the period of peak production and peak demand. The most economical method of accomplishing this is to store the energy as heated water. This substantially increases the cost of a solar energy farm.

GEOTHERMAL

The center of the earth is composed of Magma estimated to be over 9,000 degrees Fahrenheit. Closer to the surface, the temperature of the Magma decreases to 700 degrees Fahrenheit where it heats water that may reach the surface as geothermal energy in the form of steam or hot water.

Geothermal energy from hot springs, has been used for bathing and space heating since ancient times. In modern times, geothermal energy has been used to generate electricity in 24 countries and geothermal heating is in use in 70 countries.

Country	2010 Capacity MWe	Percentage of National Production
United States	3086	0.3
Philippines	1900	27.0
Indonesia	1200	3.7
Mexico	960	3.0
Italy	840	1.5
New Zealand	630	10.0
Iceland	575	30.0

Current worldwide installed capacity is 10,715 Mwe. Estimates of electricity generating potential of geothermal energy vary from 35,000 MWe to 2,000,000 MWe. The low estimate is based on present economic practicality, while the high estimate assumes new technology in drilling that will allow geothermal energy to become a commercial economic reality.

The most developed and largest geothermal field in the world is the Geysers in California, a complex of 22 geothermal power plants, drawing steam from more than 350 wells, located in the Mayacamas Mountains 72 mi north of San Francisco, California. Output from the Geysers can be considered equivalent to 3 nuclear reactors. However, safety concerns and environmental impact show geothermal energy to be the best source of energy for the future.

Developing geothermal sources by drilling into the earth's crust may produce large amounts of energy, but may also open up a "Pandora's Box" if we are not able to control the Magma that could possibly reach the surface of the earth. When drilling into the earth's crust, we must remember the environmental disaster that man caused April 20, 2010 when the Deepwater Horizon got out of control in the Gulf of Mexico.

Chevron Corporation is the world's largest producer of geothermal energy. The company's geothermal operations are primarily located in Southeast Asia where steam from underground deposits of boiling water is tapped and piped to the surface to turn generators, producing 1,273 MW of electricity.

Conclusion must be made that geothermal energy has the greatest potential for supplying the world with all its energy needs, but practical and economic considerations limit development in the near future.

NUCLEAR ENERGY

Nuclear energy could someday be our primary source of electricity if politicians, scientists and consumers understand that nuclear energy must be a compromise between economics, the environment, safety and proliferation.

This book is written in the wake of the nuclear disaster at Fukushima, Japan in 2011 which is having a profound effect on future prospects for nuclear energy. As a result, nuclear energy as it existed prior to 2011 must be considered to be an unpredictable source of energy. Prior to Fukushima, growth of nuclear energy has been declining due to economic and safety concerns. However, there is a brighter side to the nuclear issue which is held in the future of Generation IV Reactors, Pebble Bed Reactors, Fusion Technology, and possibly nuclear technology that the public is not yet aware of. Assumptions as to the future of nuclear energy would be an irresponsible position to take. Therefore, we must plan our future using renewable energy, and conserve our present sources of energy.

NUCLEAR IN ITS INFANCY

Nuclear energy to produce electricity has been developing since shortly after World War II, when on December 20, 1951, at the Experimental Breeder Reactor in Arco, Idaho, USA, electricity from the world's first nuclear power plant illuminated four light bulbs.

On June 27, 1954, the USSR's Obninsk Nuclear Power Plant became the world's first nuclear power plant to generate electricity for a power grid, and produced around 5 megawatts of electric power.

On August 27, 1956 the first commercial nuclear power plant, Calder Hall 1, England, with a net electrical output of 50 MW was connected to the national grid. The Shippingport Atomic Power

Station in Shippingport, Pennsylvania was the first commercial reactor in the USA going on-line in 1957.

World Total Electricity Production from all sources including nuclear as of 2006 was 16,000 billion kilowatt hours per year.

As of 2010 there were over 440 commercial nuclear power reactors operating in 30 countries, producing 377,000 megawatts of total capacity. They provide 14 % of the world's total electric production.

As of January 2011 there were 65 plants with an installed capacity of 63,000 megawatts under construction in 16 countries.

The following table will put nuclear energy production in perspective with other sources of electricity production.

Coal	40.8 %
Natural Gas	21.3 %
Hydro	16.2 %
Nuclear	13.4 %
Oil	5.5 %
Other	2.8 %

NUCLEAR ECONOMICS

Cost competitiveness of nuclear energy involves not just the cost of construction and operating costs, but disposal of spent fuel and permit processes. The following table compares the cost per kilowatt hour for various types of electricity production.

ELECTRICITY GENERATION	2009 SYSTEM COST PER KW HOUR
Conventional Coal	.095
Advanced Coal with CCS	.136
Natural Gas with CCS	.089
Advanced Nuclear	.114
Wind	.097
Wind Offshore	.243
Solar PV	.211
Hydro	.086

Notes to the previous table:

CCS refers to Carbon Control and Sequestration which is highly controversial and subject to change at the whim of politicians.

Natural gas is based on 2010 natural gas prices, which will increase as natural gas starts going down the depletion curve.

Hydro at the present time is the most economical, but sites for new hydroelectric dams are becoming less available, particularly in the Unites States.

Land based wind power is low in cost and is reflected in the rapid expansion of wind power farms which are growing at the rate of 30% per year.

Nuclear energy appears to be cost competitive, but public sentiment regarding safety present many unknowns which are keeping world expansion of nuclear power to a small growth rate.

Now, after a two-decade lull in construction, the U.S. is gearing up for a robust revival of nuclear power. Expanding the nuclear sector, which currently produces 20% of the nation's electricity, is considered essential to slashing carbon emissions

MAJOR NUCLEAR ACCIDENTS

The motion picture "China Syndrome", released on March 16, 1979 was a great Hollywood thriller about a potential nuclear meltdown. The plot was very realistic. Just a few days after it was released, an actual meltdown occurred at Three Mile Island. The accident at the Three Mile Island Unit 2 (TMI-2) nuclear power plant near Middletown, Pa., on March 28, 1979, was the most serious in U.S. commercial nuclear power plant history, even though it caused no deaths or injuries to plant workers or members of the nearby community. However, it changed the course of nuclear energy forever. Prior to the meltdown of the nuclear reactor at Three Mile Island, the general public was not aware of the inherent dangers of nuclear reactors. Immediately after the meltdown, public sentiment about the safety of nuclear energy combined with the economic consequences of the incident brought future development of nuclear energy to a slowdown that eventually affected the economics of the entire world.

Seven years after the Three Mile Island meltdown, the public again became aware of the inherent dangers of nuclear energy when on April 26, 1986; operators in the control room of the Chernobyl Nuclear Power Plant, in Ukrainian SSR, Reactor #4 botched a routine safety test, resulting in an explosion, and a fire that burned for 10 days. The radioactive fallout spread over tens of thousands of square miles, driving more than a quarter of a million people permanently from their homes. It remains the world's worst nuclear disaster to date. From the first day, officials downplayed the damages of the catastrophe and the politics of misinformation continues: A UN report estimates that 4,000 people will eventually succumb to cancer-related illnesses as the result of the accident. But major environmental organizations have accused the report of whitewashing Chernobyl's impact and state that more than 100,000 people have already died as a consequence of the disaster. The battle to contain the contamination and avert a greater catastrophe ultimately involved over 500,000 workers and cost an estimated 18 billion rubles, crippling the Soviet economy. Experts in the nuclear industry blame the disaster on poor design in order to reduce costs, combined with insufficient training of operating personnel. Experts also state that because there are different designs for nuclear power plants in other parts of the world, a similar disaster could not occur elsewhere.

There were no major nuclear accidents for 25 years after Chernobyl, until on March 11, 2011 when an earthquake occurred off the northeast coast of Japan causing a tsunami which inundated the multi-reactor nuclear power site in the Fukushima Prefecture of Japan. Units 4, 5 and 6 had been shut down prior to the earthquake for planned maintenance. The remaining reactors were shut down automatically after the earthquake, and the remaining decay heat of the fuel was being cooled with power from emergency generators. The subsequent destructive tsunami with waves of up to 14 meters (the reactors were designed to handle up to 5.7 meters) crippled the site, stopping the Fukushima I backup diesel generators, and caused a station blackout. The subsequent lack of cooling led to explosions and meltdowns at the Fukushima I facility, with problems at three of the six reactors and in one of the six spent fuel pools. Over the following three weeks there was evidence of partial nuclear meltdowns in units 1, 2 and 3. Radiation releases caused large evacuations, concern about food and water supplies, and treatment of nuclear workers.

These three major nuclear accidents made scientists aware that after a nuclear reactor shut down, whether manual or automatic it is necessary to have a secure source of cooling water, the ability to pump the water and foolproof means of gauging the water. The Fukushima Daiichi nuclear disaster showed that the reactor itself was adequately designed to survive the earthquake, but the simple diesel powered water pump was not adequately designed to operate during a large tsunami. It was a disaster waiting to happen.

Will future nuclear power reactors be designed and built using the costly knowledge learned from these disasters? YES.

GENERATION IV REACTORS

Generation IV reactors are a set of theoretical nuclear reactor designs currently being researched by many nations Most of these designs are generally not expected to be available for commercial construction before 2030. Current reactors in operation around the world are generally considered second or third generation systems, with most of the first generation systems having been retired some time ago. Research into these reactor types was officially started by the Generation IV International Forum which set eight technological goals, including improved nuclear safety, improved proliferation resistance, minimum waste and natural resource utilization, and decreased cost to build and run such plants.

Relative to current nuclear power plant technology, benefits for Generation IV reactors include: Nuclear waste that lasts a few centuries instead of millennia, 100 times more energy yield from the same amount of nuclear fuel, ability to consume existing nuclear waste in the production of electricity, and improved operating safety

SMALL NUCLEAR REACTORS

The photograph at left shows a testing facility that will be used to demonstrate a scaled-down version of a new modular nuclear design by Babcock & Wilcox. The tower will test the gravity-fed emergency cooling system. These nuclear reactors can generate only 125 to 140 megawatts of power, about a tenth as much as a big one. But the utilities are betting that these smaller, simpler reactors can be manufactured quickly and installed at potentially dozens of existing nuclear sites or replace coal-fired plants that may become obsolete with looming emissions restrictions.

Indeed, the smaller reactors still could incite major opposition. They face the same unresolved issues of where to put the waste and public fear of contamination, in the event of an accident. They could also raise alarms about creating possible terrorism targets in populated areas. Still, the sudden interest in small reactors illustrates a growing unease with the route that nuclear power has taken for half a century.

PEBBLE BED REACTORS

A pebble-bed reactor thus can have all of its supporting machinery fail, and the reactor will not crack, melt, explode or spew hazardous wastes. It simply goes up to a designed "idle" temperature, and stays there. In that state, the reactor vessel radiates heat, but the vessel and fuel spheres remain intact and undamaged. The machinery can be repaired or the fuel can be removed. These safety features were tested with the German AVR reactor. All the control rods were removed, and the coolant flow was halted. Afterward, the fuel balls were sampled and examined for damage and there was none.

A large advantage of the pebble bed reactor over a conventional light-water reactor is in operating at higher temperatures. The reactor can directly heat fluids for low pressure gas turbines. The high temperatures allow a turbine to extract more mechanical energy from the same amount of thermal energy; therefore, the power system uses less fuel per kilowatt-hour.

Pebble bed reactors are also capable of using fuel pebbles made from different fuels in the same basic design of reactor. Proponents claim that some kinds of pebble-bed reactors should be able to use thorium, plutonium and natural unenriched uranium, as well as the customary enriched uranium. There is a project in progress to develop pebbles from either reprocessed fuel rods or decommissioned nuclear weapons.

There is significantly less experience with production scale Pebble Bed Reactors than Light Water Reactors. As such, claims made by both proponents and detractors are more theory-based than based on practical experience.

Pebble Bed technology was first developed in Germany, but political and economic decisions were made to abandon the technology. In various forms, it is currently under development by MIT, University of California at Berkeley, the South African company PBMR, General Atomics (U.S.), the Dutch company Romawa B.V., Adams Atomic Engines, Idaho National Laboratory, and the Chinese company Huaneng.

FUSION TECHNOLOGY

Fusion power is believed to have significant safety advantages over current power stations based on nuclear fission. Fusion only takes place under very limited and controlled circumstances (by comparison fission, including catastrophic failure, only requires that there is sufficient fuel within a small enough space). For this reason, a failure of precise control or cessation of fueling quickly shuts down fusion power reactions. There is no possibility of runaway heat build-up or large-scale release of radioactivity, little or no atmospheric pollution, the power source comprises light elements in small quantities which are easily obtained and largely harmless to life, the waste products are short-lived in terms of radioactivity, and there is little overlap with nuclear weapons technology.

Waste management of a fusion reactor is more acceptable than that of a fission reactor. The radioactive inventory at shut-down may be comparable to that of a fission reactor, but there are important differences. The half-life of the radioisotopes produced by fusion tend to be less than those from fission, so that the inventory decreases more rapidly. Unlike fission reactors, whose waste remains radioactive for thousands of years, most of the radioactive material in a fusion reactor would be the reactor core itself, which would be dangerous for about 50 years, and low-level waste another 100. Although this waste will be considerably more radioactive during those 50 years than fission waste, the very short half-life makes the process very attractive, as the waste management is fairly straightforward. By 300 years the material would have the same radioactivity as coal ash.

In general terms, fusion reactors would create far less radioactive material than a fission reactor, the material it would create is less damaging biologically, and the radioactivity "burns off" within a time period that is well within existing engineering capabilities.

Fusion powered electricity generation was initially believed to be readily achievable, as fission power had been. However the extreme requirements for continuous reactions and plasma containment led to projections being extended by several decades, and more than 60 years after the first attempts, commercial power production is still believed to be unlikely before 2040.

Fusion power could solve many of the problems of fission power but, despite research having started in the 1950s, no commercial fusion reactor is expected before 2050. Many technical problems remain unsolved. Proposed fusion reactors commonly use deuterium, an isotope of hydrogen, as fuel and in most current designs also lithium. Assuming a fusion energy output equal to the current global output and that this does not increase in the future, then the known current lithium reserves would last 3000 years, lithium from sea water would last 60 million years, and a more complicated fusion process using only deuterium from sea water would have fuel for 150 billion years.

FAST BREEDER REACTORS

The Clinch River Breeder Reactor was initially conceived as a major step toward developing liquid-metal fast breeder reactor technology as a commercially viable electric power generation system in the United States. In 1971 U.S. President Richard Nixon established this technology as the nation's highest priority research and development effort. However, the Clinch River project was controversial from the start, and economic and political considerations eventually led to its demise.

Concerns about potential nuclear weapons proliferation were another serious issue for the commercial breeder reactor program, because this technology produces plutonium that potentially could be used to make nuclear weapons. Because of international concern about proliferation, in April 1977 President Jimmy Carter called for an indefinite deferral of construction of commercial breeder reactors.

Elsewhere in the world, only India, Russia, Japan and China currently have operational fast breeder reactor programs. The U.K., France and Germany have effectively shut down theirs.

FRENCH NUCLEAR PROGRAM

In 1973, just after the first oil embargo, the French government decided to rapidly expand the country's nuclear power capacity. There was little controversy over that decision. Today, France derives 75% of its electricity from nuclear energy, and France is the world's largest net exporter of electricity due to its very low cost of generation.

France has been very active in developing nuclear technology and is presently building its first Generation III reactor and planning a second. About 17% of France's electricity is from reactors using recycled nuclear fuel.

France has 58 nuclear reactors, with total capacity of over 63 GWe, supplying 421 billion kWh per year of electricity 78% of the total electricity generated there in 2011.

Should the United States and the rest of the rest of the World be paying more attention to the French nuclear programs?

UNITED STATES NUCLEAR PROGRAM

In the United States, following a 30 year period in which no new nuclear reactors were built, it is expected that 4-6 new units may come on line by 2020. Southern Company is building two nuclear reactors at Plant Vogtle near Augusta, Georgia. Cost is estimated at 14 billion dollars. Total capacity will be 2,430 MW. The construction of the two reactors has been hailed by Energy Secretary Steven Chu as "the first new nuclear power plant to break ground in decades". Southern Company is a public utility holding company of primarily electric utilities in the southern United States. It has headquarters in Atlanta, Georgia.

SECTION NINE
VISIONS OF THE 23RD CENTURY

DANGEROUS COURSE

In order to envision the 23rd century, we must track and extrapolate 21st century trends in our consumption of electrical energy and mobility fuels. From those trends, we can make responsible decisions as to the course of action to take. If we produce more energy and at the same time conserve the energy that we have, it may be possible to reach an equilibrium that will prevent armed conflicts over energy. Statistics in this book show that we are not on a course to equilibrium, but rather on a course leading to shortages that could trigger armed conflicts before reaching the 23rd century.

ENVIRONMENT

Prior to this section, there has been very little discussion about how energy effects our environment. Our environment is complex to the point that it would take a dedicated book to properly deal with our environment, particularly if we want to visualizing what our environment will be like in the 23rd century.

We know that we will have greater amounts of renewable energy in the 23rd century and we also know that renewable sources of energy will have less of an impact on our environment than fossil fuels. That's the good news. The bad news is that;

As we deplete our fossil fuels, oil will be depleted first, followed by natural gas, and finally by coal. Coal is our most abundant fossil fuel and coal causes the greatest amount of carbon pollution. There is much talk about sequestering carbon emissions, the main cause of greenhouse gases, but there has been little technical progress in actually eliminating pollution caused by coal. Therefore, health and global warming in the 23rd century will be a major concern determined by our ability to sequester carbon emissions.

THE OTHER ENVIRONMENT

Most people visualize our environment as the air we breathe, and give little attention to the environment below us which is being polluted from several sources, of which industrial waste and household garbage are of primary concern.

The author's first exposure to the problems of household waste disposal came in July 1980 at a convention of the east coast dairy industry where several representatives were sitting at a table and introduced themselves. One of the men introduced himself as William Rathje an archeologist from the University of Arizona. We all looked at him and said that he was at the wrong convention. His reply was that he is a guest speaker and that he is speaking as a "Garbologist" and that we would understand when he gave his talk. His talk started out by reminding us that archeologists dig to find out what happened in the past. He was digging when he hit an old land fill and brought up some old chicken bones. The chicken bones were from April 16, 1928 and were in such good condition that they were almost edible. First, how did he know the date? Answer is that the date was on the newspaper that the chicken bones were wrapped in. Back then we wrapped our garbage in old newspapers because it was before we had the luxury of plastic garbage bags. Second question is why the chicken bones were in such good condition? Answer was that they were buried and did not have oxygen to cause them to decompose. Third question was why Mr. Rathje was talking to us? Answer was that the dairy industry was a large contributor to the plastic found in our landfills and we had an obligation to make sure that our plastic was recycled. Contrary to efforts to recycle plastic, very little progress has been made to recycle plastic.

Plastic is manufactured from oil and pipeline liquids which will be the first of our fossil resources to be depleted. In addition to the environmental impact, plastics accounts for 4.6% of the total U.S. petroleum consumption, and accounts for 1.5% of total U.S. natural gas consumption.

By the 23rd century, plastics will be of major concern to both our environment and our consumption of depleting fossil reserves.
The good side of the picture is that plastics are readily adaptable to recycling because of the relatively low temperatures required to make them workable as compared to metals. The bad side of the

picture is that for health considerations, recycled plastics may continue to be banned from packaging of food products.

POPULATION

The 23rd century starts with the year 2200. That's less than 200 years from now, but life will be radically different from today.

Estimates of total world population at the beginning of the 23rd centuty are controversial with consensus indicating that population will probably level off. However, even with population at zero growth, energy consumption per capita will continue to increase as technology provides new means of using energy.

Planet Earth is not capable of supporting such a large population while at the same time maintaining the life style that humans were accustomed to in the 21st century. Decent living conditions for all inhabitants of the planet will be something that the politicians will talk about with no practical thoughts as to how it could be accomplished.

By the 23rd century, all fossil reserves will be well into the depletion cycle resulting in continued increases in the cost of goods and services that require fossil reserves, which will further slow our economy, which will slow the world economy, resulting in suffering and starvation.

We will depend on technological advancements to supply the world with sufficient renewable energy to prevent suffering and starvation. The first question is whether technology will supply us with enough energy and the second question is whether that energy will distributed equitably so as to prevent suffering and starvation.

If we do not find ways to stop the present trends in carbon emissions, global warming, melting of the polar ice caps, and rising sea levels, the 23rd century will see population migrating to higher ground to be safe from rising sea levels. Population migration to high ground will cause overpopulation of areas capable of accepting the new influx. The migration will cause overpopulation of inland and high ground cities. In addition to migration to large cities to seek high ground, reduced availability of mobility fuels will necessitate living and working in metropolitan areas where there will be less dependency on personal transportation.

Technology could make the above assumptions irrelevant if we develop practical alternatives to replace our dependency on fossil reserves.

DISTRIBUTION OF WEALTH

The 21st century witnessed an increase in the gap between the wealthy and the middle class. Americans assumed that the gap was the result of hard work. However, when the housing bubble burst in the year 2008, Americans learned how a few speculators in the financial industry made vast sums of money by using unscrupulous tactics causing the poor to lose their homes, and the world to go into recession.

During the 2012 United States presidential campaign, wealth of the candidates became an issue for the first time in political history. Candidates showed that equitable taxation was an issue, with the wealthy being taxed at lower rates because their wealth came from capital gains while the working class paid higher taxes because they did not have money to invest.

As we approach the 23rd century, the link between energy and wealth will become more apparent as the wealthy squander energy resources while the poor cannot afford the cost of transportation to get to work or heat their homes.

LEADERSHIP

The sequence of events leading to the 23rd century will be - technology – communications – revolution – leadership – equality.

The year 2011 has seen modern communications as a weapon of revolution that is transforming oppressive governments into democratic governments with leaders elected by the consent of the people. The time span between the year 2011 and the 23rd century will allow the voting public to learn, with the help of modern communications, how to select the best qualified leaders to guide their future.

Just as oppressive governments have been transformed into democratic governments, existing democratic governments such as the United States must be transformed into progressive democratic governments. The United States government was formed in 1776 having a constitution that was based on 2.5 million people riding horse and buggies. There was no thought that the same government structure of 1776 would be ruling over 350 million people riding in jet airplanes with small communications devices that enable them to know instantaneously what is happening with their neighbors next door as well as their neighbors half way around the world.

In the 20th century the United States became the most powerful country in the world. By the 21st century we learned that foreign governments were controlling the way we lived and did business. Example is the 1973 oil embargo that revealed our dependence on foreign governments for essential oil necessary for every day existence. Our government took action that was too little and too late.

Some radical people would like to tear up the constitution and form some new type of government. Those people do not understand that the constitution has provision for change if original parts of the constitution did not anticipate what the future would be like. Our problem is not with the constitution, but rather with the people that we elect and expect to rule in accordance with the intentions of the constitution.

On April 18, 1977, President Jimmy Carter made a speech to the nation alerting us to the impending energy crisis. The nation did not want to hear about an energy crisis. Politicians claim that the speech was the reason that Jimmy Carter was not reelected to a second term. His speech on energy policy is included in section three of this book. If he gave that same speech today, would the nation listen? Probably not. If he could give that speech when we enter the 23rd century, the nation will ask why they were not told about a looming energy crisis sooner.

It will be a very difficult task to convince people that they should take action now to prevent an energy crisis that may not occur until the 23rd century, 200 years from now. Unless radical changes are made in the way the United States governs itself, particularly when it comes to energy reserves, we will find that foreign governments are controlling our everyday lives.

Leaders are expected to learn from their mistakes and from the mistakes of others. Our leadership let us down by allowing the Dot-Com bubble to inflate and then burst. Our leadership let us down by allowing the housing bubble to inflate and then burst. Our leadership let us down when they allowed the largest Ponzi scheme in U.S. history to inflate and then burst. Lack of regulation and oversight allowed these bubbles to inflate and eventually burst. Will our elected leaders allow lobbyists determine how and when our corporations and financial institutions are regulated?

On the other side of the coin, leaders are expected to learn from their previous successes and the successes of other nations. An example is the path to energy independence that France is pursuing. With little fossil fuel resources, France will probably become energy independent by the 23rd century, while the United States continues to talk about energy independence but does very little to achieve it.

Rationing has been referred to as the action of last resort, only when depletion of fossil fuels affects the economy of the United States and the world. Wake up politicians, it has.

RATIONING

Gasoline was rationed during World War II. It was accepted as necessary to fight the war. Again, it was considered during the 1973 oil embargo when members of the Organization of Arab Petroleum Exporting Countries (OAPEC) proclaimed an oil embargo against the United States. Energy Czar at that time was William E. Simon who ruled that gasoline rationing would only be a last resort. Rationing of transportation fuels was never mentioned after 1973 because of United States political considerations. It has always been considered the third rail of politics, because touching the issue meant instant political death. Few voters would elect a candidate who proposed gasoline rationing.

Rationing of mobility fuels by the 23rd century is inevitable. It will be the action of last resort when there is not enough fuel to fill up our cars, and long lines for gasoline reminiscent of 1973 return.

Rationing of fossil fuels which is becoming more apparent in the 21st century can be made more effective and equitable if advanced planning is used to work out the details. It will not be a simple task because there are many greedy and selfish individuals who will try to find ways to circumvent the system and profit from it. There will be lobbyists from many industries trying to exempt their jet planes and yachts from rationing. Rationing must be an equitable system that will discourage gas guzzling cars, private boats, and private planes. It must be an equitable system that will prevent the wealthy from squandering fuel while the middle class does not have enough fuel to get to work.

Rationing must be planned well in advance of its actual need to prevent squandering and to provide for the many essential services such as police, fire, ambulances, food production, and the military. Implementation of rationing will be based on electronic scanning systems would have to be designed and distributed to the many different priorities.

Leaders must plan for the worst and hope that it will not be necessary to put their plans into action.

There will be a good side to rationing. New fuels such as Algae Fuels would not be rationed, allowing their prices to rise above gasoline made from fossil reserves. The middle and upper classes would pay more for Algae Fuel for their large gas guzzling cars, yachts and jet airplanes thereby injecting more money into new industries allowing them to construct larger more efficient facilities for renewable energy.

TECHNOLOGY

Technology is impossible to predict. An example of the uncertainty is trying to predict the type of fuel that commercial aircraft will use in the 23rd century. In a report by Boeing, January 23, 2006 titled "Alternate Fuelled Aircraft", the disadvantages of Ethanol and Liquid Hydrogen as aircraft fuel were discussed. There was no mention of fuel made from Algae. Less than 6 years later on November 8, 2011 a Boeing 737-800 flew from Houston, Texas to Chicago Illinois on Algae Fuel. At the beginning of the 21st century, it is difficult to predict the practical possibility of replacing mobility fuels presently produced from fossil reserves with mobility fuels produced from Algae. If Algae Fuel becomes practical, Ethanol Fuel will be phased out and the controversy over food for fuel will no longer be an embarrassment in the United States.

At the beginning of the 20th century, electricity produced from nuclear reactors was considered to be the answer to the World's energy needs well into the future. These hopes were stymied by three nuclear accidents, waste disposal problems and economic considerations. However, research continues for safe nuclear energy with fusion technology possible by the 23rd century.

The 23rd century and beyond will see a shift in World sources of energy from depleting fossil fuels to electricity produced from renewable sources. Nuclear energy is considered to be a renewable source of energy capable of providing the World with enough electric energy for thousands of years because of the abundance of nuclear fuel. However, nuclear energy will remain controversial because of the safety factor, particularly apparent after the nuclear disaster caused by the Fukushima, Japan tsunami. However, nuclear energy could provide us with enough electricity if technological advances allow safe, environmentally friendly energy from

The 20th century has seen rapid growth in electricity produced from wind power, with growth rates of 30% annually. At this rate, electricity generated from wind power will surpass nuclear energy by the year 2020. However, the inherent problem of wind power is the need for sites with sufficient wind velocity to produce adequate electrical energy. Therefore, lack of sites for large wind farms may limit future development.

Technology will provide us with new sources of renewable energy, but the world will never have unlimited energy that can be squandered by selfish individuals and corporations. We must learn to live with limitations on our consumption of energy or otherwise there will be shortages resulting in suffering, starvation and eventually armed revolution.

TRANSPORTATION

The greatest energy challenge of the 23rd century will be in the transportation sector, particularly for commercial airplanes and ships that requires mobility fuels.

Predominant use of energy for ground transportation will depend upon new technological developments for electricity production by safe nuclear power or possibly wind power. As electricity becomes available in sufficient quantity, it will phase out mobility fuels produced from fossil reserves. At the beginning of the 21st century, it appeared that our ground transportation would emerge utilizing;

Batteries will be used for short range vehicles such as personal commuting, short distance delivery trucks, work and construction vehicles. Increased use of battery powered vehicles will depend upon breakthroughs in the cost, safety and life expectancy of batteries.

Overhead trolley wires will continue to be used and will be increased for rail and inner-city trolleys. Present systems will be expanded for long distance freight trains and high speed passenger trains.

Electromagnetic induction systems similar to monorail trains at amusement parks and some high speed rail will be expanded to cover more inner-city transit vehicles and eventually could cover highway vehicle propulsion.

Electromagnetic induction for commuter trains is being developed by Bombardier, a major manufacturer of transit vehicles and systems. They recently announced and demonstrated a new power system for light rail vehicles, that doesn't use overhead wires. The power system transfers energy by electromagnetic induction, making direct electrical connections unnecessary.

Electromagnetic induction works by using a varying electric current (or magnetic field) to induce a current in a nearby loop of wire. Basically, it allows for electric energy transfer without physical contact between the wires. That's important because it allows the track-side electrical equipment to be covered and out of the weather, and also completely insulated so it doesn't pose a hazard to pedestrians or cyclists. The power-sending coil is embedded under the road surface, and a receiving coil in the train collects power. The system proposed by Bombardier utilizes rails to keep the vehicle aligned with the magnetic induction wires.   Future systems will have an optical guidance strip in the road that the vehicle will automatically follow to keep the magnetic induction wires precisely aligned. When there is a temporary interruption in the electromagnetic induction wires resulting from a crossroad or bridge, the vehicle will switch to battery power.

The next prediction will raise a lot of eyebrows. By the 23rd century, embedded electromagnetic induction and navigation will allow the driver to catch some sleep. Signals on the navigation system will alert the driver to changes in system requiring attention. If the driver does not intervene, automatic breaking will take over and stop the vehicle. Obstructions such as other vehicles or animals in the path will be sensed by simple radar on the vehicle which will automatically bring it to a safe stop. Automatic breaking systems were introduced by Mercedes at the beginning of the 21st century and have been proved practical and reliable.

The 23rd century will see tractors equipped for electromagnetic induction propulsion used on interstate highways to pull trailers, or double trailers, or possibly triple trailers. When it is time to leave the interstate highway, the trailers will be hooked up to an electric battery propelled tractor for the short trip to its final destination.

Gasoline and diesel produced from oil will be too expensive for individual transportation, being reserved for essential services such as law enforcement, fire, ambulances and the military. As we approach the 23rd century rationing of fossil reserves will allow less and less for individual usage making us more dependent on renewable energy. One of the promising sources of mobility fuel will be Algae Fuel. As of the year 2012 Algae Fuel was in its infancy and difficult to predict how much will be available for future needs.

Gas guzzling vehicles will be replaced by small cars, two wheeled vehicles and bicycles. Gasoline will be replaced by renewable fuels such as Ethanol and Algae Fuel while electric cars replace gasoline power for short commutes.

Mass transportation will become more acceptable as commuters find new ways to make productive use of their time on mass transport.

Ocean transport of commercial goods will become increasingly expensive as oil becomes depleted. Shipping companies and governments have recognized this dilemma. Shipping companies are minimizing the problem of fuel consumption by ordering larger ships with better hull designs, operating at slow speed and using slippery bottom paints. Eventually, the cost of fuel will result in less ocean transportation and deterioration of world economic trade. With this in mind, shipping companies looked at the successful operation of military ships using nuclear propulsion. It seemed a normal course of action to consider nuclear energy for commercial shipping. However, attempts by four nations, United States, Germany, Russia and Japan to power commercial ships with nuclear energy have been discontinued. Discouraging results are discussed in section eight of this book.

Cruise ships offer passengers the opportunity to visit foreign countries at a price that they could not otherwise afford. The largest single expense operating a cruise ship is the fuel. As oil reserves become depleted, the cost of fuel paid by the cruise lines will increase necessitating higher prices for tickets. Unless the cruise lines change their operating procedures, there will be a decrease in passengers. One method of reducing fuel consumption is for the cruise ships to travel shorter distances at much lower speeds. However, because passengers want to see the greatest number of ports on a single cruise, competition will prevent "slow" cruising. Fuel conservation can be substantially improved if international governments intervene or the cruise lines agree among themselves to fuel saving strategies. Cruise ships, together with commercial ships and aircraft will increasingly depend on renewable mobility fuels to replace oil which will be well into depletion by the 23rd century.

Future mobility fuels such as ethanol and hydrogen have not been proven to be practical leaving algae fuel as the only mobility fuel that can propel our ships and planes. The question is whether algae fuel can be produced in sufficient quantities to supply the world needs.

LIFESTYLE CHANGES

One of the easiest ways to visualize the 23rd century is to take the 20th and 21st centuries and extrapolate. The fast food industry coined the expression "super-size" for their meals. The public went for it. Retailing introduced super-size food and super markets. The public went for it and the markets grew bigger and bigger. Shopping malls grew from 50 stores in the first shopping malls to 1,500 stores in the South China Mall in Dongguan, China. The bad news is that if there is a super-size shopping center nearby, the mom and pop store is out. The good news is that it will not be necessary to drive the gas guzzler between the dry-cleaners and the theater.

The cruise industry started with 1,000 passenger ships and has grown to ships carrying 5,000 passengers. By the 23rd century they may carry 10,000 passengers. They may be powered by small nuclear reactors or may fill up with 300 tons of algae fuel. Only technology will determine how they perform.

The Mayo Clinic is an example of how medicine has evolved from a doctor making house calls to a public clinic serving three metropolitan areas with total staff of over 56,000. Size has made them more efficient and technically competent. This trend in medical care should continue well into the 23rd century, if our politicians see this as an example of what free enterprise can do.

Riyadh, Saudi Arabia boasts the largest airport. The King Khalid International Airport is a massive 484-square miles. It is an example of what western oil money can buy.

Dubai, in the United Arab Emirates boasts the world's tallest building, the Burj Khalifa 2,717 feet high with 163 floors. Dubai also boasts its International Airport Terminal 3 with the second largest amount of floor space of 12.76 million square feet. Another example of what western oil money can buy.

Mecca, a city in Saudi Arabia hosts the building with the largest floor space. The Abraj Al Bait Towers has 16.15 million square feet of floor space. Guess where the money came from.

Electronic business started in the 20th century and will continue to be refined through the 23rd century. Executives will be conducting business electronically reducing the need for transportation. Teleconferencing, email and FAX will replace flying long distances to conduct business. Salesmen will depend upon web pages to sell their products.

Manufacturing businesses will grow and become more consolidated providing housing to workers rather than employ workers who would have to commute long distances to reach their jobs.

The 23rd century will see continuation of competition for retail sales between super-size shopping centers and on-line shopping. Technology will be the determining factor. If algae fuel is available in sufficient quantities, the consumer will drive to the shopping center in fuel guzzling vehicles. If mass transportation is made attractive to the consumer, they will leave the fuel guzzling vehicle at home and enjoy the benefits of the super-size shopping malls.

In the United States, bicycles are used for exercise and recreation. Few use their bicycles as an alternate means of transportation the way bicycles are used in European and Asian countries. The 23rd century will see more bicycle paths for recreation such as the

East Coast Greenway together with more bicycle paths for commuters. The United States will learn from Europe.

CITIES

If sufficient energy is made available from nuclear energy and algae fuel, cities of the 23rd century will continue to be built horizontally with individual private homes spread out over many square miles utilizing personal transportation rather than mass transportation. However if technology does not provide us with sufficient energy, cities will consolidate into more compact fuel efficient complexes many stories high with mass transportation used within the city.

New cities will be built from scratch with even the smallest cities centering on super-sized shopping malls which in many cases will be housed in enormous domed structures to allow efficient cooling and heating. The trend toward consolidated health facilities offering all levels of care from family medicine to assisted living will grow in size and will be housed in compact fuel efficient enclosures. This trend towards super-sizing will include sports complexes, entertainment centers and educational facilities. The exception to centralized centers will be eating establishments which must be accessible immediately when people get hungry.

Population will continue to grow well into the 23rd century resulting in many cities growing in vulnerable areas such as flood zones, earthquake faults, nuclear sites and coastal areas susceptible to tsunami disasters.

If the controversy over global warming proves to be correct and melting of the ice caps continues at a rate that results in rising sea level of only a few feet, it will be necessary for entire cities to relocate to higher ground causing rapid expansion of existing cities. We may not know until the 23rd century, the true extent of global warming. We know that global warming will continue as long as we depend on fossil fuels that emit carbon dioxide. Technological advances in nuclear energy and algae fuel may be the only practical answer to global warming

EMPLOYMENT

Employment is determined by the need for goods and services, consistent with the ability to produce those goods and services. There will always be a need for more and more goods and services, but there may not be the ability to produce those goods and services because of the increased cost of energy, both electrical energy and mobility fuels.

Housing starts declined rapidly after the 2008 housing bubble burst because of the financial repercussions causing an excess of foreclosed homes on the market. At the same time, cost of energy to produce the components for new homes was increasing at historic rates, thus further widening the gap between existing homes on the market and the cost of building new homes. The housing market further stagnated causing further unemployment.

Recreational travel by car, ship and plane slowed because of both the increased cost of fuel for transportation and a reduced market caused by unemployment, thus causing further unemployment.

Essential goods such as food increased in price because of increases in the fuel which is necessary to produce food. Consumers spent more of their money on food, leaving less of their income to spend on housing and recreational travel.

It is becoming more obvious that energy is the common denominator for the production of consumer goods. With less fossil fuel available for energy and byproducts such as plastics and medicine, there will be less employment in the consumer products industries.

This logic can go on and on showing the domino effect caused by increased cost of fuel resulting from depletion of fossil reserves. Now try to visualize an absurd situation where a renewable fuel such as algae fuel becomes available at zero cost. Cost of essential goods, recreational travel and other services will decrease resulting in greater employment. Now try to visualize a more realistic situation where renewable bio-fuels are produced by labor that would otherwise not be utilized. The results could be very positive if politicians put their differences aside and develop work projects to produce renewable bio-fuels.

If we cannot produce sufficient renewable energy from safe nuclear reactors and bio-fuels, the shift in employment will be from skilled jobs in energy intensive industries to lower paying jobs in the service industries. Such shift in jobs will increase the gap between the wealthy and the middle class thus further increasing the possibility of revolt and revolution in civilized nations.

COMMUNICATIONS

Prior to the 19th century, and the development of the electrical telegraph, physical communication, generally by horse or ship were the means of delivering messages from one person to another. Electrical communications developed rapidly in the 20th and 21st centuries allowing communications between all parts of earth and even beyond. Mankind was no longer dependent on mobility to communicate.

The 23rd century will see electronic communications opening new sources of information to the general public that in the past had been deliberately hidden by both democratic governments and dictatorships. Telephones, cell phones, television, internet and iPods will do much more than just inform the public about the weather. Information will become available regarding politics, repression, wealth, how wealth was acquired, tangible assets, intangible assets, taxes paid and how wealth is being squandered. The middle class will learn how corporations, wealthy people and government officials are squandering resources while the middle class are being deprived of essential goods and services.

The beginning of the 21st century and particularly the year 2011 saw advanced communications as the rallying point of revolutions around the world and in the United States.

REVOLUTION

The year 2011 saw peaceful revolutions in the United States such as the Tea Party and Occupy Wall Street. Elsewhere in the world there were armed rebellions against tyrannical dictatorships, some of which are;

2011 Egyptian revolution
2011 Libyan revolution
2011 Syrian uprising
2010–2011 Algerian protests
2011 Bahraini uprising
2011 Iraqi protests
2011 Jordanian protests
2011 Moroccan protests
2011 Omani protests
2011 Yemeni uprising
2012 Syria uprising

These trends show that most countries in Middle East and Africa are vulnerable. Is there revolution in the air? How fast will it spread? How far will it spread by the 23rd century? Will peaceful rebellions in the United States become violent? Will there be armed conflicts between nations because of energy? Will there be internal armed rebellions because of energy?

Answer to all these questions can be resolved peacefully if we learn that we are in a new age of communications where there can be no secrets between people, nations and leaders. Inequalities will be exposed to all and communication tools of rebellion will be used to achieve what rightfully belongs to all.

These rebellions all had in common, access to communications which motivated and inspired repressed people to take action. Will these rebellions spread if our limited energy resources are not shared equitably? YES.

OIL AS A WEAPON IN WAR

Oil is the lifeblood of modern civilization. Individuals became aware of this when the oil embargo of 1973 exposing the vulnerability of U.S. energy supplies, setting off an unprecedented energy crisis. Before that time, Americans had become accustomed to using energy without concern about where the oil came from. In 1973 expectations about energy supply changed as gasoline prices skyrocketed to levels previously thought impossible. The 1973 oil embargo was a diplomatic war with oil as the weapon and the United States lost the war by succumbing to the demands of a foreign nation in order to assure a continuing supply of essential oil.

Japan attacked Pearl Harbor on December 7, 1941. The main reason for the attack was to gain the ability to acquire raw materials and oil to support their imperialistic expansion. The US and Brittan had successfully isolated them from virtually all of their materials, particularly oil, necessary to support their expansion. Japan received petroleum from three sources: The USA, Indonesia and Burma. The USA movement included all three sources. Japan would not accept a withdrawal from the Chinese war and instead began planning a first strike against the USA navy. Eliminating or reducing the USA naval forces in the Pacific would make the Japanese navy paramount, and thus Japan would be able to defeat the economic consequences of the USA ultimatum. After eliminating the USA navy Japan planned to occupy the Dutch East Indies and Burma, thus gaining control of enough oil to run their military and economy.

Prior to the start of aggressions by Germany leading to World War II, both Germany and Britain knew that oil would be essential if Germany attempted to dominate the world. The reliance of Germany on oil and oil products for its war machine was identified before the war. Strategic bombing started with RAF attacks on Germany in 1940. Part of the immense Allied strategic bombing effort during the war, included refineries for natural oil, plants producing synthetic oil, storage depots, and other petroleum, oil and lubrication products. World War II, by far the greatest conflict ever fought, dramatically emphasized the indispensability of oil to any strategy for victory. Some historians take the position that World War II was the first war decided on oil supply.

Iraq invaded Kuwait because they needed the money that Kuwait oil would provide. Western countries were largely of the opinion that Iraq's invasion of Kuwait was largely motivated by its desire to take control over the latter's vast oil reserves. Some feel there were several reasons for the Iraqi move, including Iraq's inability to pay more than $80 billion that had been borrowed to finance the Iran-Iraq war and Kuwaiti overproduction of petroleum which kept revenues down for Iraq. Why did the US intervene on the side of Kuwait? OIL

Again in 2012, oil was used as a weapon when Iran threatened to interrupt world oil supply by closing the Strait of Hormuz if there was any interference in their attempt to develop nuclear weapons. Iran's ambassador to the United Nations said closing the Strait of Hormuz, the passageway for about a fifth of the world's oil trade, is an option if his country's security is endangered.

Will the United States fight for Middle East for oil? YES. The addict will fight to feed his addiction. If Iran attempts to close the Strait of Hormuz, there is no question that the United States will fight to keep the supply of oil flowing. This is one more example of going to the brink of war for oil.

ECTION TEN
SUMMARY

Population of the world is increasing at an alarming rate and will continue to do so well into the predictable future. Increases in population combined with increasing energy consumption per capita are placing an untenable burden upon our fragile planet. The beginning of the 21st century emerges having enormous problems with population, energy and economics. Controlling population growth at this late stage may not solve our energy problems. The result of excessive population could be starvation, suffering and economic recession. Government is the only way of controlling population and politicians consider meddling with population to be the third rail of politic, touch the subject and you are politically dead.

On April 18, 1977, former president of the United States, Jimmy Carter gave a speech warning us of the looming energy crisis. Americans did not want to hear about a future energy crisis. He put solar panels on the roof of the White House. President Ronald Regan removed the solar panels. Leadership did not act on the looming energy crisis until it was starting to show adverse effects on our economy. When our government started to take action, it was too little and too late.

Energy is the common denominator of all goods and services. The nations with the greatest reserves of fossil fuels are, or will become the wealthiest nations for as long as their reserves are in demand. We have seen this happening with Middle East oil producing nations. The reverse of this will be nations such as the United States who consume more energy than they produce resulting in trade deficit which will ultimately increase national debt with associated economic consequences. As the world depletes its resources of fossil fuels, attempts are being made to stretch the remaining fossil fuels and to create more sustainable sources of energy. However, progress in creating sustainable energy is not keeping up with increasing demands which will eventually result in shortages of everything from food and water to durable consumer goods.

Planet Earth is consuming fossil reserves at a rate that will result in <u>oil going into the depletion phase</u> about 2020, natural gas going into its depletion phase in less than 100 years and coal going into its depletion phase in less than 200 years. United States entered into its oil depletion phase in 1970, as shown by oil production reaching a peak at that time. Depletion is defined as a resource reaching peak production and then becoming uneconomical for further extraction of the resource. The cost of extracting those minerals from the earth will become increasingly expensive to the point where they can no longer be extracted from the earth. This particularly applies to oil that in the 21st century has become more difficult, dangerous, and expensive to extract from the earth. Soon after oil reaches the point of being impractical to extract from the earth, coal and natural gas will follow.

Scientists, intellectuals, statisticians and educators understand that we will deplete our finite amount of fossil reserves in 50, 100 or 500 years from now. <u>Who cares?</u> Certainly not the oil company executives who want to make their fortunes now. Not the politicians whose only concern is to be reelected. Not the consumers who only care about the price of gasoline at the pump today.

The 20th century witnessed the <u>depletion of a vast source of oil</u> from Alaska's Prudhoe Bay Oil Field on the North Slope. In 1974 it was estimated to have 10 billion barrels of oil. Production was started in 1977. In 1988 at its peak, production was 2.0 million barrels a day. In 2010, production was down to 600,000 barrels per day. The Energy Department said that an optimistic assumption of remaining reserves in the North Slope through 2050 is 36 billion barrels of oil. That would only be enough to meet current U.S. oil demand for about five years, without importing any oil, but some major obstacles stand in the way of meeting those estimates. New sources of oil are available from shale oil within the United States, which are more costly to produce and present numerous environmental concerns. It is too early to determine how shale oil will affect the World's potential production of oil.

Energy has <u>two main categories</u>. First is electrical energy used to power anything that can connect to the electrical grid system. The other is mobility fuels for cars, trucks, busses, planes and ships that cannot connect to the electrical grid system. Mobility energy comes from liquid fuels refined from fossil reserves, liquid fuels from sustainable sources, and batteries that can be charged from electricity. Electricity can be produced from numerous sources ranging from our finite fossil reserves which will eventually be depleted, to renewable sources such as hydropower and wind energy.

As fossil fuels, particularly oil, become depleted, sustainable energy will become the primary source of fuel for vehicles that cannot be plugged into the electrical grid causing the cost of mobility fuels to increase to a level that will adversely affect world economies.

<u>Technology</u> is providing us with new ways of producing fossil energy by means of fracking, as well as sustainable energy utilizing algae.. However, technology is also providing us with new ways of consuming energy that will increase demand faster than we can develop new sources of energy.

The <u>United States consumes 23%</u> of the World's oil production. Of this, United States cars and light trucks consume 10.5% of the World's oil production which is more than the total oil consumption of China, or Japan and India combined, or Russia and Brazil and Germany combined. As of 2010 the United States imported more than half of its oil from foreign countries, some of which are enemies of the United States. The good news is that larger amounts of oil will be imported from newly discovered oil sands in Canada, reducing our imports from hostile nations.

The government of the United States recognized the looming energy crisis in mobility fuels. Actions emphasized Ethanol as a mobility fuel resulting in the <u>controversy of food for fuel</u>. Ethanol is not practical for our ships and planes and is not the answer for our need for mobility fuels. However, there is a possibility that fuel derived from Algae will prove to be practical as a mobility fuel. Ships and planes have tested Algae fuel, with positive results. However, as of 2011, the cost of Algae fuel was considerably more expensive than fuel derived from fossil reserves. With sufficient time, technology and ingenuity, Algae fuel may be the answer to our mobility fuel needs.

Ethanol was the first attempt by politicians to provide a sustainable fuel for transportation vehicles. It became controversial because it consumed large amounts of electrical energy and land for crops that led to the phrase "food for fuel". It resulted in increases in the cost of most categories of food.

Algae fuels are being developed that could provide a sustainable source of fuel for transportation vehicles, requiring less land for feedstock as compared to Ethanol. Algae fuels can utilize a wide variety of plant-based sugars such as sugarcane based sucrose, corn-based dextrose, and sugar from other sources including cellulosics. April 2012, Solozyme Inc. and Bunge Inc. formed a joint venture to produce sustainable triglyceride oils for use in fuel applications in the Brazilian market. The companies expect startup in the second half of 2013 with production of 100,000 metric tons of oil. Other pilot plants are being built within the United States to produce Algae Fuel on a smaller scale. Department of Energy released a study in 2011 showing that there is enough feedstock for Algae in the United States to produce 85 billion gallons of biofuels, enough to replace 30 percent of the nation's petroleum consumption.

Sustainable electrical energy from wind power had growth from year 2000 to year 2007 averaging over 30% per year. As of 2007 wind energy provided over 3% of the World's total energy requirements. Growth at the rate of 30% cannot continue indefinitely due to the exhaustion of sites that are practical for new wind farms. If it were possible to maintain growth rate of 30% per year for wing energy, by the year 2030 electrical output would be equivalent to that of hydro power.

Politicians and industrial leaders are reluctant to introduce drastic reform such as rationing, tax reform and population controls. If action is not taken in a timely manner, the remaining energy resources of the world will be left for the wealthy to purchase and squander. Modern communications will reveal how the middle and lower classes are being deprived while the wealthy are left to squander the remaining resources. Modern communications will also provide a conduit for the deprived to unite, with results that could threaten world economics. The following paragraphs are not directly related to our energy crisis, but rather are intended to show that our leaders are not capable of foreseeing looming problems, and after they occur, are not capable of correcting the underlying cause of the problem.

The United States was warned about the Dot-Com bubble by former Chairman of the Federal Reserve, Alan Greenspan. In a speech given at the American Enterprise Institute during the Dot-com bubble of the 1990s, he coined the phrase "Irrational exuberance" which was a warning that the market might be overvalued. The government did not act. Those who had money invested in the stock market saw their retirement dreams disappear. Leadership did not act to prevent the Dot-Com bubble from bursting, and never took later action to prevent a similar "correction" from destroying the financial future of Americans.

The United States was warned about the dangers of the multi-trillion-dollar derivatives market whose crash helped trigger the financial collapse in the fall of 2008. Brooksley Born, who from an obscure government position campaigned to have the derivatives market regulated. Opposing her were the nation's top financial experts, former Fed Chairman Alan Greenspan, former SEC Chairman Arthur Levitt, and Treasury Secretary Robert Rubin. Greenspan, Levitt and Rubin prevailed and no action was taken to regulate the derivatives market. After the collapse of the housing market, Greenspan admitted that his opposition to regulating the derivatives market was "clearly a mistake." Knowing that derivatives were the underlying cause of the 2008 recession, neither the United States government nor the Securities and Exchange Commission ever took action to regulate the derivatives market.

By the spring of 2008, burdened by billions of dollars of bad mortgages, the Wall Street investment banks began to go bankrupt. Collapse of the investment bank industry was a prelude to the risk management meltdown of Wall Street in September 2008, and the subsequent global financial crisis and recession. The United States poured billions of dollars into the "Too Big to Fail" investment banks and insurance companies with little accountability.

We <u>cannot depend upon politicians</u> to resolve our energy problems. We can only hope that government will back off and allow the private sector to develop new sources of fossil energy and sustainable energy without interfering by mandating price supports, incentives, import duties and taxes.

SECTION ELEVEN
SOURCES OF DATA

U. S. ENERGY INFORMATION ADMINISTRATION provides many statistics used in this book. It reports to The U. S. Secretary of Energy, which is a political appointee of the President of the United States. The United States Department of Energy is a Cabinet-level department of the United States government concerned with the United States' policies regarding energy. The United States Department of Energy is administered by the United States Secretary of Energy.

The United States Department of Energy Organization Act of 1977 established the Energy Information Administration (EIA) as the primary Federal Government authority on energy statistics and analysis, building upon systems and organizations first established in 1974 following the oil market disruption of 1973.

The United States Department of Energy is located in the building shown above at 1000 Independence Avenue, Washington, DC 20585.

EIA is an organization of about 380 Federal employees, with an annual budget in Fiscal Year 2010 of $111 million. EIA is the Nation's premier source of energy information and, by law, its data, analyses, and forecasts are independent of approval by any other officer or employee of the United States Government.

By law, EIA's products are prepared independently of policy consid-erations. EIA neither formulates nor advocates any policy conclusions. The Department of Energy Organization Act allows EIA's processes and products to be independent from review by Executive Branch officials; specifically, Section 205(d) says: "The Administrator shall not be re-quired to obtain the approval of any other officer or employee of the Department in connection with the collection or analysis of any infor-mation; nor shall the Administrator be required, prior to publication, to obtain the approval of any other officer or employee of the United States with respect to the substance of any statistical or forecasting technical reports which he has prepared in accordance with law."

The United States Energy Information Administration prepares and presents data on when, where, how much energy, and what kind of energy. The amount of data handled by the EIA can be overwhelming. Presentations are made in tabulations, graphs and charts which are available to the general public. Industrial and government leaders depend upon data from the EIA to make decisions that will affect both the United States and the world. Historical data presented by the EIA is generally based on accurate records of corporations and governments leaving little room for doubt as to their accuracy. However, projections are based on theoretical means that may be questionable.

NATIONMASTER.COM is an Encyclopedia of Statistics covering a massive central data source and a handy way to graphically compare nations. NationMaster is a vast compilation of data from such sources as the CIA World Factbook, UN, and OECD. It includes a vast number of subjects from Agriculture to Transportation. Caution should be taken because some of the data may be up to 5 years old.

WIKIPEDIA the Free Encyclopedia has numerous subjects and is constantly increasing in scope. Some subjects are current on news headlines. Editing is controversial but appears well controlled and acceptable.

U.S. CENSUS BUREAU *Statistical Abstract of the United States*, published since 1878, is the authoritative and comprehensive summary of statistics on the social, political, and economic organization of the United States. Use the Abstract as a convenient volume for statistical reference and as a guide to sources of more information both in print and on the Web. Sources of data include the Census Bureau, Bureau of Labor Statistics, Bureau of Economic Analysis, and many other Federal agencies and private organizations.

U.S. DEPARTMENT OF COMMERCE BUREAU OF ECONOMIC ANALYSIS produces economic accounts statistics that enable government and business decision-makers, researchers, and the American public to follow and understand the performance of the Nation's economy. To do this, BEA collects source data, conducts research and analysis, develops and implements estimation methodologies, and disseminates statistics to the public.

U.S. GOVERNMENT ACCOUNTABILITY OFFICE (GAO www.gao.gov) is an independent, nonpartisan agency that works for Congress. Often called the "congressional watchdog," GAO investigates how the federal government spends taxpayer dollars.

UNITED NATIONS STATISTICS DIVISION is committed to the advancement of the global statistical system. We compile and disseminate global statistical information, develop standards and norms for statistical activities, and support countries' efforts to strengthen their national statistical systems. We facilitate the coordination of international statistical activities and support the functioning of the UN Statistical Commission as the apex entity of the global statistical system.

U.S. CENTRAL INTELLIGENCE AGENCY. The World Factbook provides information on the history, people, government, economy, geography, communications, transportation, military, and transnational issues for 266 world entities. Publishes and updates a directory of *Chiefs of State and Cabinet Members of Foreign Governments* regularly. The directory is intended to be used primarily as a reference aid and includes as many governments of the world as is considered practical, some of them not officially recognized by the United States.

WORLD NUCLEAR ASSOCIATION supports the global nuclear energy industry with expert Working Groups focused on industry goals and concerns. The WNA Public Information Service, is the world's leading resource for facts on nuclear energy.

WTRG ECONOMICS gathers data from industry, government and proprietary sources to bring the most reliable and current information to bear in addressing the needs of our clients. Using years of experience together with sophisticated economic, statistical and financial analysis tools we are able to assist clients in analyzing their company's condition and in anticipating the direction in which their industry and company is headed.

FREE TRADE MAGAZINES (freetrademagazines.com) provides a free magazine subscription service to qualified professionals in all industries. The trade publications on the website benefit subscribers by providing them with the most up-to-date information they need to run their businesses more effectively.

ONLINECONVERSION.COM converts just about anything to anything. Has over 5,000 units, and 50,000 conversions. Most popular conversion pages are Length, Temperature, Speed, Volume, Weight, Cooking, Area, Fuel Economy, Currency. The appendix of this book contains pertinent conversion factors.

THE U.S. DEPARTMENT OF ENERGY'S OFFICE of Energy Efficiency and Renewable Energy has developed the Power Technologies Energy Data Book to provide a current and accurate set of comprehensive power technology-related data. The data book is an evolving document and is updated periodically.

THE NATIONAL ACADEMIES PRESS (NAP) was created by the National Academies to publish the reports issued by the National Academy of Sciences, the National Academy of Engineering, the Institute of Medicine, and the National Research Council, all operating under a charter granted by the Congress of the United States. The NAP publishes more than 200 books a year on a wide range of topics in science, engineering, and health, capturing the most authoritative views on important issues in science and health policy.

GEOTHERMAL ENERGY ASSOCIATION is a trade association composed of U.S. companies who support the expanded use of geothermal energy and are developing geothermal Resources worldwide for electrical power generation and direct-heat uses. The GEA advocates for public policies that will promote the development and utilization of geothermal resources, provides a forum for the industry to discuss issues and problems, encourages research and development to improve geothermal technologies, presents industry views to governmental organizations, provides assistance for the export of geothermal goods and services, compiles statistical data about the geothermal industry, and conducts education and outreach projects.

UNION OF CONCERNED SCIENTISTS is the leading science-based nonprofit organization working for a healthy environment and a safer world. UCS combines independent scientific research and citizen action to develop innovative, practical solutions and to secure responsible changes in government policy, corporate practices, and consumer choices. What began in 1969 is now an alliance of more than 250,000 citizens and scientists. UCS members are people from all walks of life: parents and businesspeople, biologists and physicists, teachers and students. Achievements over the decades show that thoughtful action based on the best available science can help safeguard our future and the future of our planet.

SECTION TWELVE
DOOM AND GLOOM BOOKS

United States politicians have been aware of the impending energy crisis for many years and have done little to change our demand for energy. However, ordinary citizens have realized the situation and have written numerous books on the subject. Here are a few:

A PRESIDENTIAL ENERGY POLICY, Michael C. Ruppert

Simple but widely ignored concepts relating to the critical role of oil and gas in the modern world. First, they are finite resources, formed in the geological past, therefore subject to depletion. Second, they have to be found before they can be produced. Many claims have been made that new technology will counter the natural decline, but there is an irony: the better the technology, the faster the depletion. the impact on farming and population, and the nature of Money. The impact on the economy is a central theme of the book. It gives emphasis to the U.S. situation but also covers the wider World, ending with twenty-five sensible recommendations by which the United States Government could react to the unfolding situation.

ACTING IN TIME ON ENERGY POLICY, Kelly Sims Gallagher

Policymakers tend to do just enough to satisfy political demands but not enough to solve the real problems, and they wait too long to act. The resulting policies are overly reactive, enacted once damage is already done, and they are too often incomplete, incoherent, and ineffectual. This important volume details this problem, making clear the unfortunate results of such short-sighted thinking, and it proposes measures to overcome this counterproductive tendency.

BLOOD AND OIL, Michael T. Klare

In 1998, for the first time, more than half of the United States' oil supply came from foreign sources. Imports of foreign petroleum are projected to rise to 60% in 2010, 65% in 2020, and 70% by 2030. America's oil fields are experiencing irreversible decline. Even if reserves in Alaska's wildlife refuge are extracted, they will reduce U.S. imports of foreign oil by only about 2% per year for the next two decades-an almost negligible change. To pay for all of this imported oil, American businesses and consumers will have to cough up an estimated $3.5 trillion between 2000 and 2025- if oil remains within a moderate price range. Even more importantly, since most of the world's remaining oil resources are located in deeply unstable regions, over 65% in the Persian Gulf alone, military involvement to ensure access is inevitable.

GUSHER OF LIES: THE DANGEROUS DELUSIONS OF ENERGY INDEPENDENCE, Robert Bryce

In this controversial, meticulously researched book, Robert Bryce exposes the false promises and political posturing behind the rhetoric. *Gusher of Lies* explains why the idea of energy independence appeals to voters while also showing that renewable sources like wind and solar cannot meet America's growing energy demand. Along the way, Bryce exposes the ethanol scam as one of the longest-running robberies ever perpetrated on American taxpayers. In a new foreword to this edition, he shows how energy independence rhetoric was used during the 2008 election, even as the heavily subsidized ethanol business fueled a grow- ing global food crisis.

OIL PANIC AND THE GLOBAL CRISIS: PREDICTIONS AND MYTHS, Steven M. Gorelick

Are we running out of oil or do we have plenty of this resource? Will the oil age end before we run out of oil? Have we reached the maximum daily production rate of oil or are we already on the down slide? Profes- sor Gorelick has compiled the necessary data and provided his own incisive analysis to assist the reader in understanding the complex issues related to the supply and demand hydrocarbons.

OVER A BARREL: THE COSTS OF U.S. FOREIGN OIL DEPENDENCE, John Duffield

Making the obvious but often overlooked point that depending on imported oil carries more than economic consequences at the gas pump and the home furnace, Duffield notes the costs to American consumers, such as skyrocketing heating bills from government foreign policy and military efforts to protect unreliable overseas supplies. So far, those policy responses have increased rather than decreased costs. ...Although Duffield is dubious about American intervention overseas, he does endorse American hegemony as a route to changing oil-related attitudes and policies worldwide.

THE COMING OIL CRISIS, C.J. Campbell

The history and current status of the important oil industry are re- viewed in this study of the geological origins of oil and gas. Assessed are how much oil and gas has been produced, what remains in known fields, and what is yet to be found, with attention to how to properly interpret published numbers, many of which are false or distorted by vested interests. The contention is made that the growing Middle East control of the market is likely to lead to a radical and permanent increase in the price of oil before physical shortages begin to appear within the first decade of the 21st century. The book further argues that the coming oil crisis will create economic and political discontinuity of historic proportions, as the world adjusts to a new energy environment.

THE EMPTY TANK: OIL, GAS, HOT AIR, AND THE COMING GLOBAL FINANCIAL CATASTROPHE, Jeremy Leggett
Argues incisively that oil production has peaked, with dwindling supplies and soaring prices in the offing. Worse than the possibility that the world cannot cope, he feels, is the threat that it will cope all too well by burning more coal and coal-derived synthetic fuels, thus exacerbating atmospheric carbon dioxide build-up and global warming. This will lead the planet down "the road to horror," illustrated in a sketchy montage of the usual environmental doomsday scenarios: rising sea levels, extreme weather, famine and war

THE END OF OIL: ON THE EDGE OF A PERILOUS NEW WORLD, Paul Roberts
All economic activity is rooted in the energy economy, which means a substantial portion of the current world economy is linked to the production and distribution of oil. But what will happen, Roberts asks, when the well starts to run dry? Walking readers through the modern energy economy, he suggests that grim prospect may not be as far off as we'd like to think and points out
proposal for the last, best, and most practical global response to Peak Oil.

THE RACE FOR WHAT'S LEFT, Michael T. Klare
The world is facing an unprecedented how political unrest could disrupt the world's oil supply with disastrous results. *The Race for What's Left* takes us from the Arctic to war zones to deep ocean floors, from a Russian submarine planting the country's flag under the North Pole to the large-scale buying up of African farmland by Saudi Arabia and other nations.

THE GOLDEN CENTURY OF OIL 1950-2050: THE DEPLETION OF A RESOURCE, C.J. Campbell
A French geologist presents the facts and figures for the nonspecialist reader on when and where the oil will run out. An overall assessment includes a review of how oil gets down in the ground, how it gets back

up, how we can (and can't) measure it, and how much is left. The regional assessment discusses reserves and production for nine major areas. Then each country and giant oilfield is detailed in graphs and tables.

THE LAST HOURS OF ANCIENT SUNLIGHT: Revised and Updated: The Fate of the World and What We Can Do Before It's Too Late by Thom Hartmann

THE MYTH OF THE OIL CRISIS: OVERCOMING THE CHALLENGES OF DEPLETION, GEOPOLITICS, AND GLOBAL WARMING, Robin M. Mills

Oil has been blamed on everything from triggering global climate change to being the catalyst of geopolitical wars. The result of which has led to calls for developing alternative energy sources such a nuclear, solar and wind, as well as creating more efficient vehicles. Mr. Mills discusses in a concise and easy understandable manner the various viewpoints concerning oil production, such as the Geologists, the Economists, the Militarists, the Environmentalists, coal, biofuels, gas liquids, and potential alternatives such as wind, solar, and nuclear. The Myth of the Oil Crisis is an excellent book for those seeking a better understanding of our dependence on oil, how we got that way, and what we can do about.

THE OIL DEPLETION PROTOCOL, Richard Heinberg and Colin Campbell

A Plan to Avert Oil Wars, Terrorism and Economic Collapse. The challenge of peak oil without the protocol, a century of chaos? Dealing with diminishing oil, options and strategies. A unique crisis of resource depletion—a crisis that goes beyond "peak oil" to encompass shortages of coal and natural gas, copper and cobalt, water and arable land. With all of the Earth's habitable areas already in use, the desperate hunt for supplies has now reached the final frontiers.

WHEN OIL PEAKED, Kenneth S. Deffeyes

On a time scale somewhere between one hundred and three hundred years, our civilization has to come around to sustainable and renewable resources. Most energy will be, directly or indirectly, solar." Offering the admittedly unpopular alternatives of uranium and coal until that happens, he discusses means of minimizing dangers and reducing energy consumption; his comparison of the efficiency of various forms of transportation may make readers think again about barges coming from China

AN INCONVENIENT TRUTH: THE CRISIS OF GLOBAL WARMING, Vice President Al Gore

Popular, controversial and great photos. As the result of this book, action has been taken to reduce carbon emissions and has made the world more aware of the effects of burning fossil fuels.

POPULATION BOMB, POPULATION CONTROL OR RACE TO OBLIVION, Paul R. Ehrlick 1968.

He begins the work that gave him instant notoriety (infamy) by saying: "I have understood the population explosion intellectually for a long time." It isn't simply that his predictions turned out to be wrong in some of the particulars, but rather that they were so completely wrong that they will NEVER come to pass (though he unrepentantly continues to beat the same drum).

Sometimes, a book is too far ahead of the times. This is one of those books. Population is one of the causes of our present energy crisis and the cause of starvation in many underdeveloped countries. What would life be like in China or India if the population of those countries was only 500,000?

THE PARTY'S OVER: OIL, WAR AND THE FATE OF INDUSTRIAL SOCIETIES, by Richard Heinberg

CROSSING THE RUBICON: THE DECLINE OF THE AMERICAN EMPIRE AT THE END OF THE AGE OF OIL, by Michael C. Ruppert

THE END OF OIL: THE UNMAKING OF AMERICA'S ENVIRONMENT, SOCIETY, AND INDEPENDENCE, by Michael J. Graetz

THE PRIZE: THE EPIC QUEST FOR OIL, MONEY& POWER, by Daniel Yergin

OIL: MONEY, POLITICS AND POWER IN THE 21ST CENTURY, by Tom Bower

THE HALLIBURTON AGENDA: THE POLITICS OF OIL AND MONEY, by Dan Briody

CONVERSION FACTORS

1 thousand thousand	1 million (6 zeros)
1 thousand million	1 billion (9 zeros)
1 thousand billion	1 trillion (12 zeros)
1 thousand trillion	1 quadrillion (15 zeros)
1 thousand quadrillion	1 quintillion (18 zeros)
1 barrel petroleum	42 gallons
1 gallon	3.785 liters
1 gallon water at 20 C	8.330 pounds
1 gallon jet fuel	6.793 pounds
1 gallon auto gasoline	6.294 pounds
1 gallon diesel fuel	7.000 pounds
1 gallon residual fuel oil	7.549 pounds
1 barrel ethanol feedstock	5.930 million BTU
1 barrel diesel	5.825 million BTU
1 barrel crude oil	5.800 million BTU
1 barrel jet fuel	5.661 million BTU
1 barrel biodiesel feedstock	5.433 million BTU
1 barrel biodiesel	5.359 million BTU
1 barrel conventional gasoline	5.253 million BTU
1 barrel propane	3.836 million BTU
1 barrel ethanol fuel	3.563 million BTU
1 barrel LPG	3.558 million BTU
1 million BTU	7.200 gallons diesel
1 million BTU	7.240 gallons crude oil
1 million BTU	7.420 gallons jet fuel
1 million BTU	7.995 gallons gasoline
1 million BTU	11.788 gallons ethanol
1 million BTU	.975 cu. Ft. natural gas
1 cubic meter	35.314 cubic feet
1 cubic foot natural gas	1.026×10^3 BTU
1 cubic meter natural gas	36.232×10^3 BTU

1 billion cubic meter natural gas	36.232 x 10^{12} BTU
1 cube 10x10x10 natural gas	1.026 x 10^6 BTU
1 short ton coal	2,000 pounds
1 long ton coal	2,240 pounds
1 short ton coal	20.2 billion BTU
1 calorie	.00397 BTU
1 calorie (nutritional)	3.968 BTU
1 watt hour	3.412 BTU
1 kilowatt hour (KWh)	3.412 x 10^3 BTU
1 thousand kilowatt hours	3.412 x 10^6 BTU
1 megawatt hour (MWh)	3.412 x 10^6 BTU
1 million kilowatt hours	3.412 x 10^9 BTU
1 gigawatt hour (GWh)	3.412 x 10^9 BTU
1 billion kilowatt hours	3.412 x 10^{12} BTU
1 trillionwatt hour (TWh)	3.412 x 10^{12} BTU
1 trillion kilowatt hours	3.412 x 10^{15} BTU
1 quadrillionwatt hour (QWh)	3.412 x 10^{15} BTU
1 kilowatt for 1 year	29.889 x 10^6 BTU
1 megawatt for 1 year	29.889 x 10^9 BTU
1 (GW) billionwatt for 1 year	29.889 x 10^{12} BTU
1 (TW) trillionwatt for 1 year	29.889 x 10^{15} BTU
1 kilowatt for 1 year	8,760 KWh per year
1 megawatt for 1 year	8,760 MWh per year
1 (GW) billionwatt for 1 year	8,760 GWh per year
1 MWe	1 x 10^6 watts (6 zeros)
1 BWe	1 x 10^9 watts (9 zeros)
1 TWe	1 x 10^{12} watts (12 zeros)
1 kilogram	2.205 pounds
1 pound	.4536 kilogram
1 meter per second	2.237 miles per hour
1 mile per hour	.447 meter per second